LC4A and the Big Test:

A Math-Story Planet Workbook

LC4A and the Big Test:

A Math-Story Planet Workbook

K. Sherrerd Lowry, Ph.D.

With Illustrations by the Author

DUER PRESS

ISBN-13: 978-0-692-40893-3
ISBN-10: 0-692-40893-2

Published by Duer Press
www.duerpress.com
Printed in the United States of America

DEDICATION

To my mother for her support,

and to my students for what they've taught me.

Contents

LC4A Wakes Up

LC4A sat up in her sleep-pod. Her mother's voice came from the next room.

"Hurry, LC! Your big math test is today! You have to catch your spiral tube to school!"

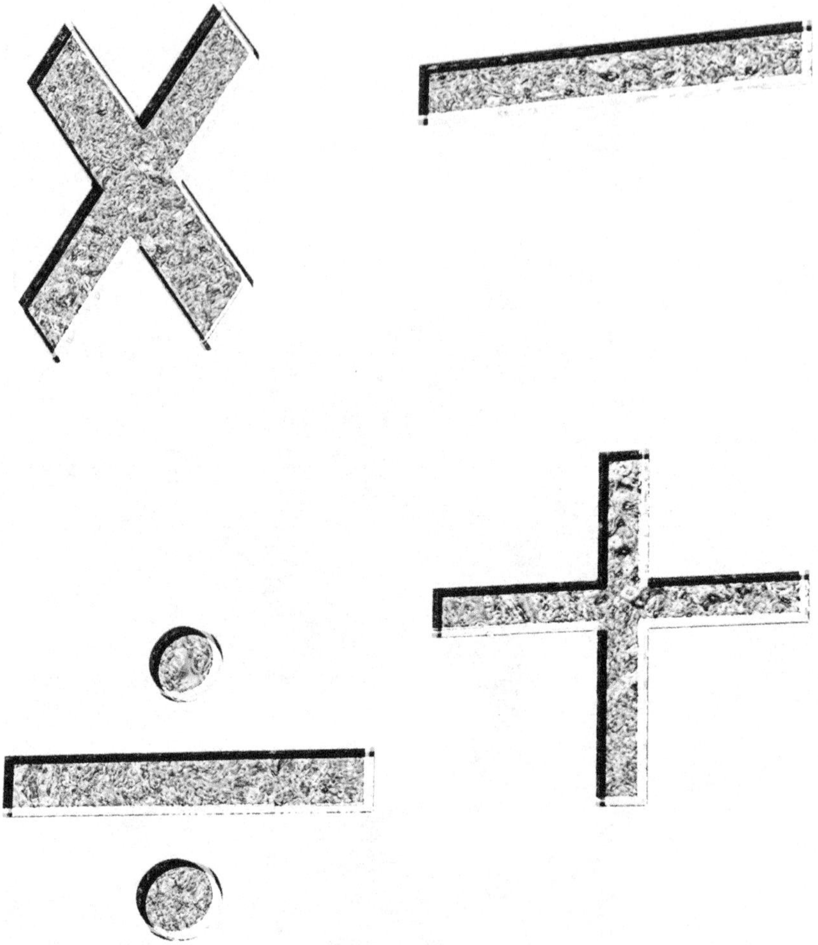

LC jumped out of her pod. It was true! Her big math test was today!

This was going to be a hard test. But LC had studied. She wanted to do her best.

It was getting late! LC needed to hurry, to catch her spiral tube to school.

1

2

3

LC caught her spiral tube, and got to class just in time.

4

5

6

Mr. D8N, the teacher, greeted LC. "Hurry to your seat, LC4A. The test is about to begin!"

7

8

The children quickly found their seats. "Remember to think your answers into your gel cubes," said Mr. D8N. "Look for the keys to choosing an operation."

As Mr. D8N started up his Problem Projector, LC tried to stay calm. "I know this," she thought to herself. "I know the keys to look for. The keys will help me solve these problems!" The class held their breath for Problem 1.

9

10

Problem #1:

#1:

In a pond, 21 eye-fish swim. 15 are red and the rest are blue. How many

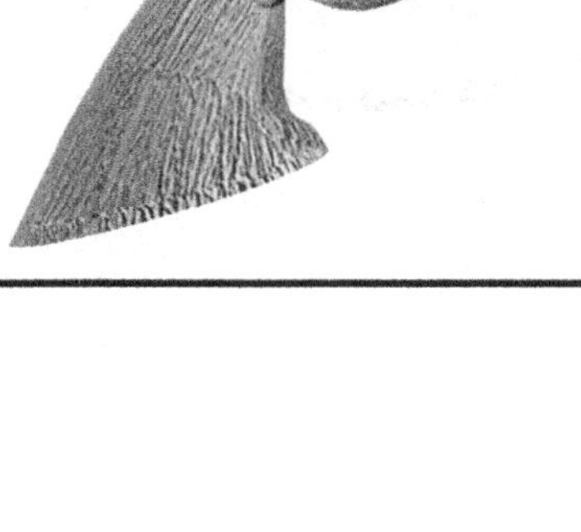

eye-fish are blue?

11

12

Solution to Problem #1:

LC4A looked at the words of the problem: **In a pond, 21 eye-fish swim. 15 are red and the rest are blue. How many eye-fish are blue?** LC decided to use counters on paper to model the problem.

She put down 15 red counters, in 3 rows of 5. Then she counted on from 15 to 21, drawing a circle each time. When she got to 21, she stopped and counted her circles. There were 6 of them!

To check, she put a blue counter down on each circle she had drawn. 6 blue counters. Then she re-counted all the red (15) and the blue (6). Yes, it added up to 21.

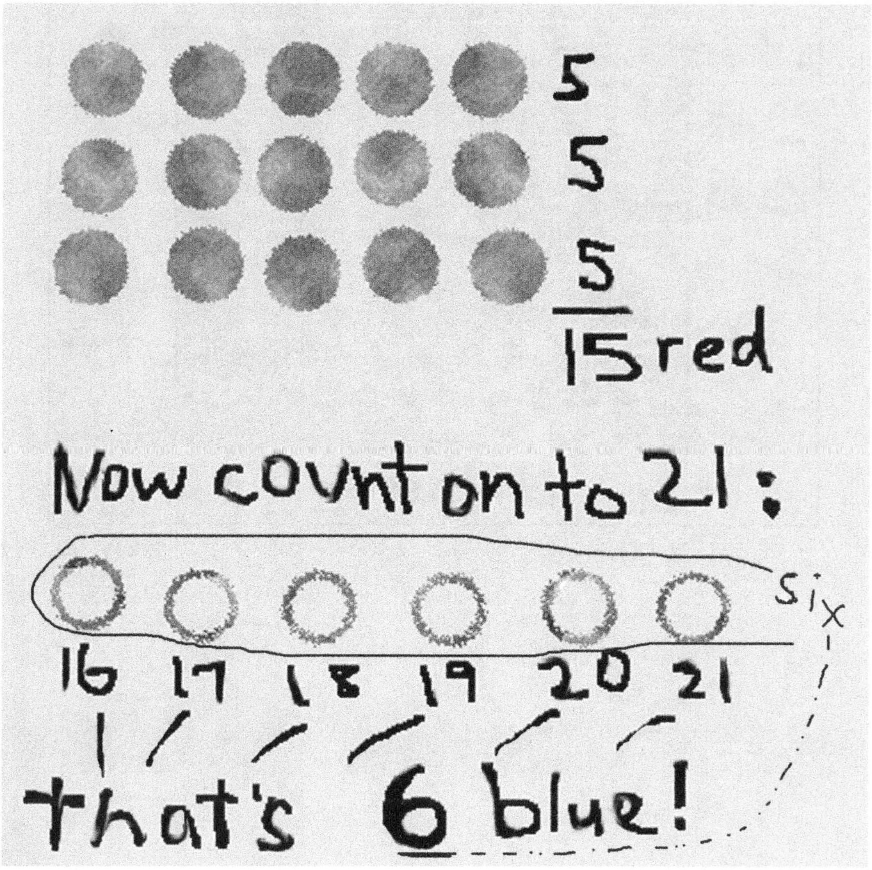

This is the equation: 15 + __ = 21

LC knew that when you have 21 altogether, and 15 are red, and the rest are blue, you can find out how many are blue either by counting on, or by subtracting: 21 - 15 = __

Modeling with counters is a good way to see what the problem is about. Either way, the answer is 6 blue fish.

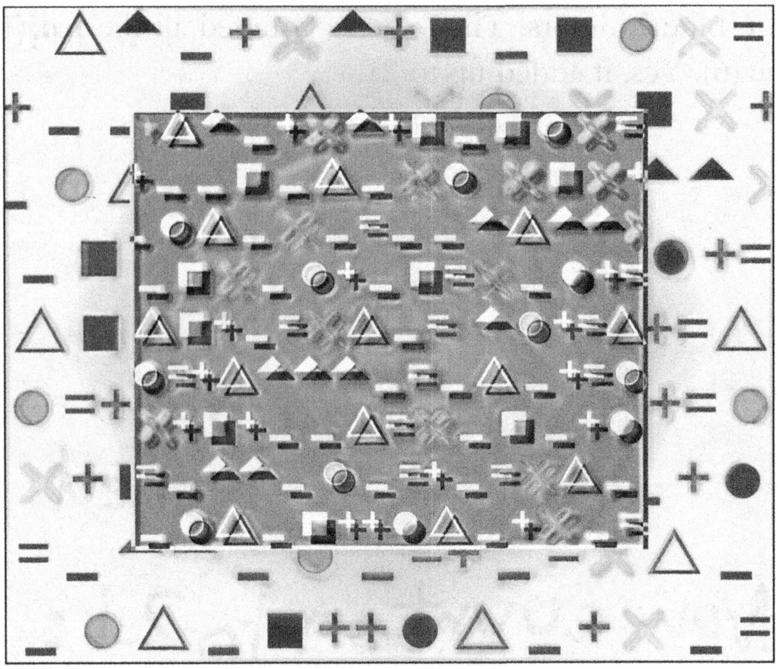

Problem #2:

#2

12 pwerts sit on a nup-tree. Some fly off. Now there are 5 pwerts on the nup-tree. How many flew off?

13

14

15

Solution to Problem #2:

LC read the problem again: **12 pwerts sit on a nup-tree. Some fly off. Now there are 5 pwerts on the nup-tree. How many flew off?**

This was like Problem #1 in some ways. But different, too. LC decided to model the problem to see what was happening. That way she could decide which operation to use.

This time LC started with 12 counters, and a drawing of a tree. That was all the 12 pwerts on the tree in the beginning. Now some flew away. She could move counters to show them flying off the tree. But how many to move? LC decided to put her hand on 5 of the counters to hold them in place, then move the rest off the tree. It looked like this:

That was the answer: 7 flew off. The equation was: 12 - __ = 5. Another way to think of it was 12 - 5 = __

The answer was the same, either way. LC remembered what Mr. D8N always said: model the problem to see what's going on!

Problem #3:

#3:

Some fly-boards are in the air. 4 more join them. Now there are 16 fly-boards in the air. How many were in the air to begin with?

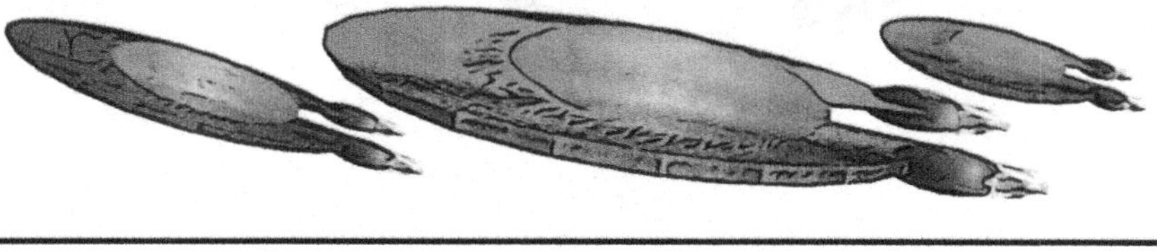

18

19

Solution to Problem #3:

LC thought about the problem: **Some fly-boards are in the air. 4 more join them. Now there are 16 fly-boards in the air. How many were in the air to begin with?**

What was happening? LC knew how many boards were in the air at the end. She decided to draw a picture of what was happening. That would help her understand what operation she should use.

LC drew all the boards from the end of the problem. There were 16, the problem said. She drew them in four rows of four each. It looked like this:

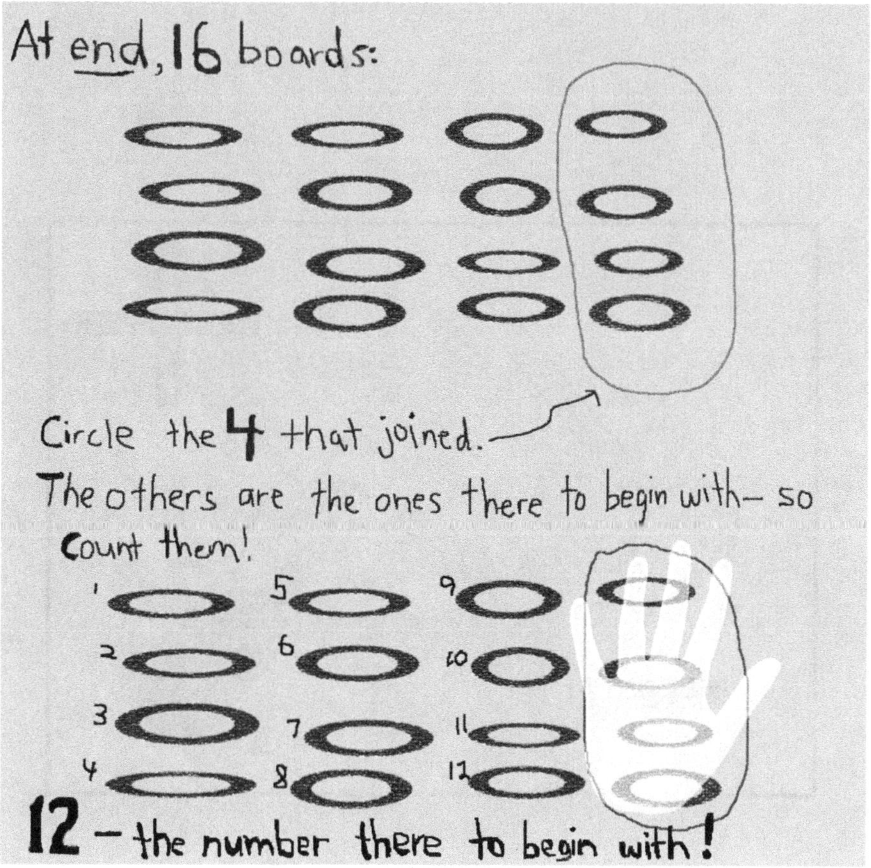

In red, LC circled the 4 boards that joined. She knew that "joined" meant that they were added to the boards already in the air.

So the boards other than those 4, were the boards already in the air in the beginning. All she had to do was count them!

She counted and found that there were 12. LC checked and found that the 12, added to the 4 that joined, did add up to the 16 fly-boards that were there in the end. LC saw that this could be shown in an equation: ___ + 4 = 16

Another way to show the same thing happening: 16 - 4 = ___. LC saw the key for this problem. It was to think about the total number of boards in the air at the end (16). Then, take away the 4 that joined. The number she got, 12, was the number of boards in the air to begin with!
Finding the key made solving the problem easy!

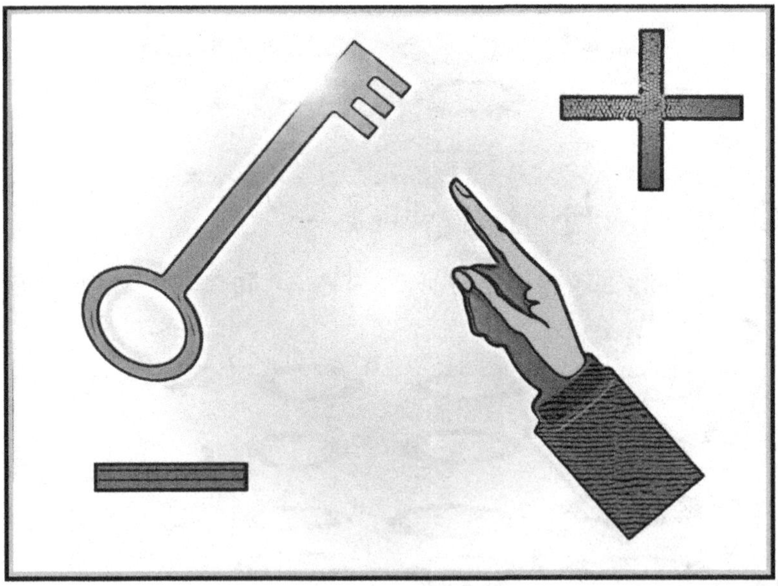

Problem #4:

#4:

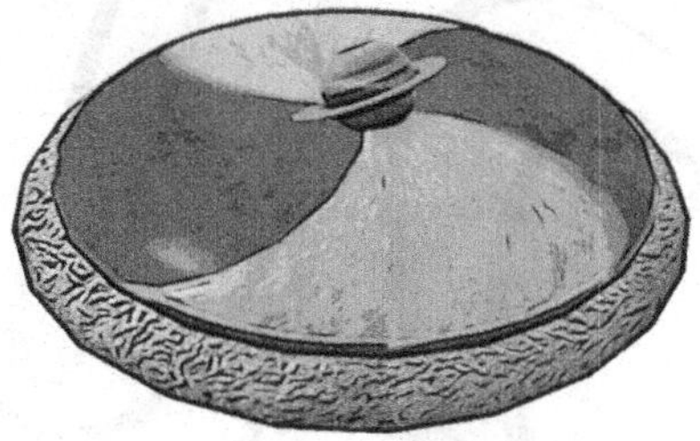

PT has 25 treats for the class party. CC has 34 treats for the party. How many more treats does CC have?

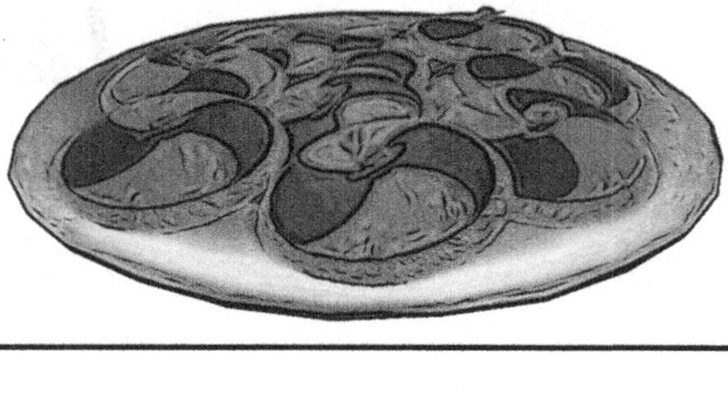

20

21

22

Solution to Problem #4:

LC looked at the problem. She looked for the key to solving it:
PT has 25 treats for the class party. CC has 34 treats for the party. How many more treats does CC have?
LC saw that she needed to compare the two numbers, 25 and 34. She decided to use a number line to model those two numbers. It looked like this:

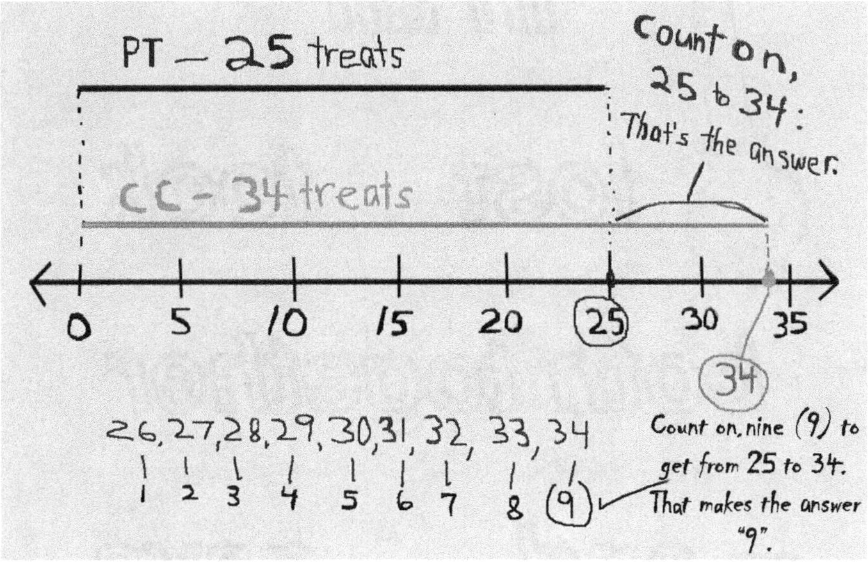

LC counted on from 25 to 34. She had to jump nine numbers on the number line to get from 25 to 34. That was the answer! CC had 9 more treats for the party than PT had.
The equation looked like this: 25 + ___ = 34

LC saw that the same answer came from subtracting:
34 - 25 = ___ Why? Because comparing means deciding how much bigger or smaller. These were some key words that tell you to subtract: "How many more?" ... "How much bigger?" ... "How much smaller?" ... "How many fewer?" ... "How many less?" ... "Compare." LC knew she had found the key for this problem!

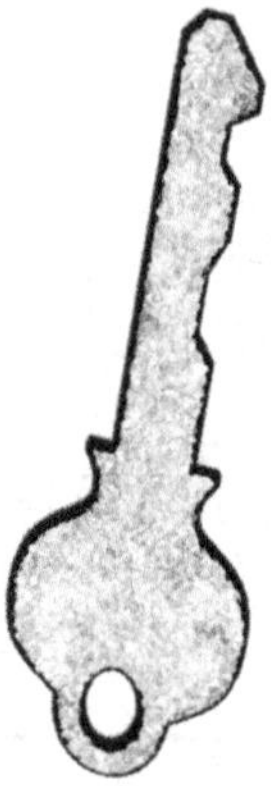

in all

lost took

both together

and gave

sold

Problem #5:

#5:

The Planets Farm

has 8 more zilk-zorks than the Stars Farm. The Planets Farm has 22 zilk-zorks. How many does the Stars Farm have?

more than

less than

sum

Solution to Problem #5:

LC read the problem: **The Planets Farm has 8 more zilk-zorks than the Stars Farm. The Planets Farm has 22 zilk-zorks. How many does the Stars Farm have?**

This was hard to figure out. LC knew two things about the Planets Farm: that it had 22 zilk-zorks in all, and that 22 was 8 more than the number of zilk-zorks on the Stars Farm.

What should she do first? What operation should she use?
LC remembered that Mr. D8N always told the class: You can always RE-WRITE the problem with smaller numbers. Sometimes that makes it easier to know what is happening. That will let you see which operation to use.

LC decided to use the numbers 3 and 7 in place of the numbers 8 and 22. She'd re-write the problem exactly, but with the new numbers. Here is what she wrote:

The Planets Farm has 3 more zilk-zorks than the Stars Farm. The Planets Farm has 7 zilk-zorks. How many does the Stars Farm have? Now it was easier to draw a picture of the problem.

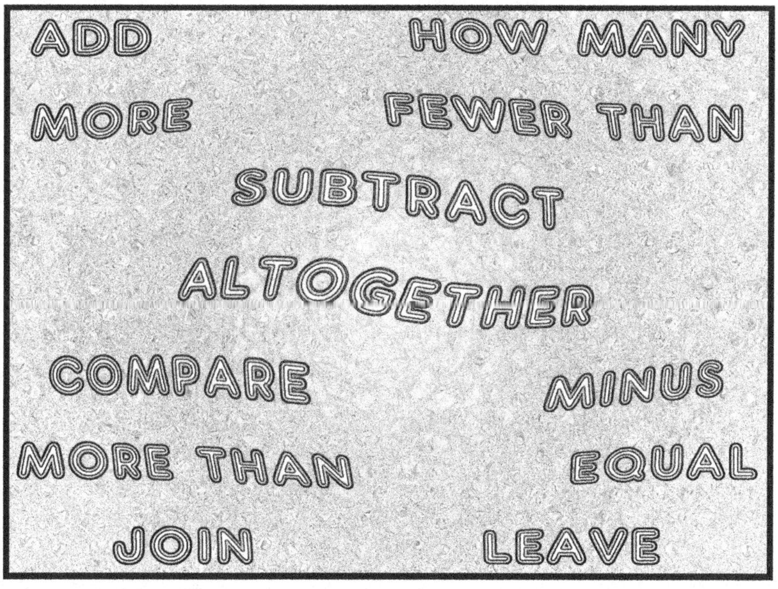

On the next page is LC's drawing of the "smaller numbers" problem situation:

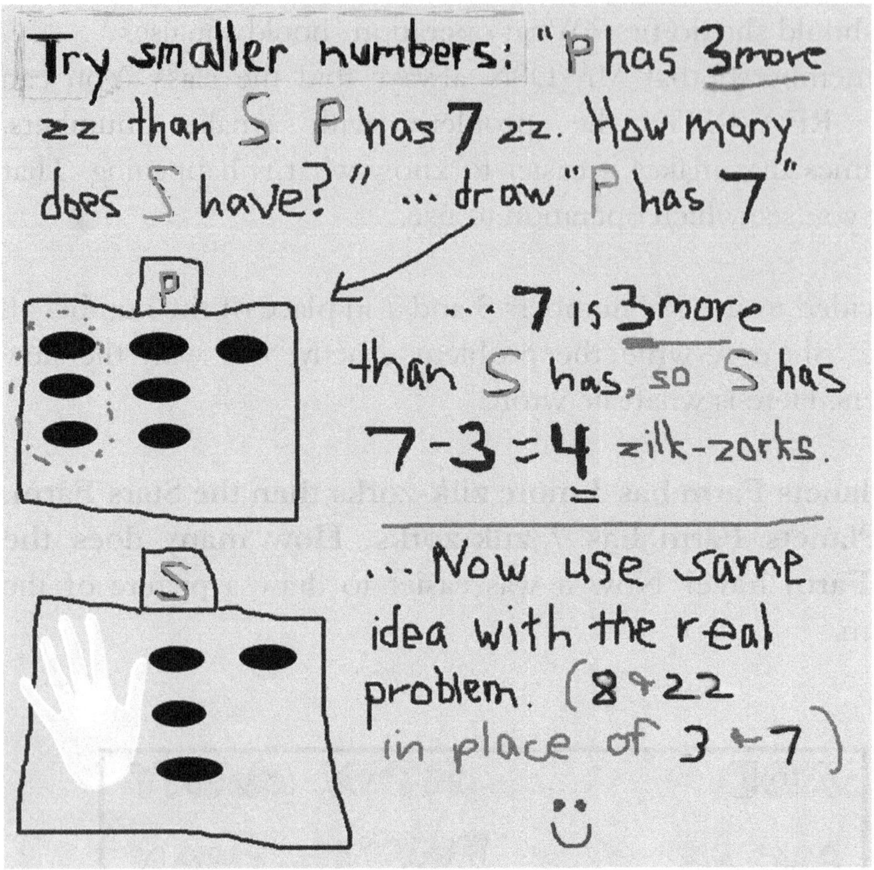

The key in the “smaller numbers” problem: the "3 more" which meant LC should subtract. Since 7 is “3 more” than the Stars Farm number, then the right operation must be 7 – 3 = 4.
This meant that in the real problem, the Planets Farm 22 minus the "8 more" would give the number of zilk-zorks on the Stars Farm. LC should subtract 8 from 22.

22 (the Planets Farm number) is ___ (the Stars Farm number) plus 8 more. Equation: 22 = ___ + 8

This means the same thing as 22 - 8 = ___. 22 – 8 = 14. So, the Stars Farm has 14 zilk-zorks. LC checked: yes, 14 plus 8 more does equal 22. She'd found the key!

Problem #6:

#6:

11 yellow toy cars are on the table. 17 blue cars and 19 red cars are also on the table. How many toy cars in all are on the table?

23

24

Solution to Problem #6:

LC saw that there were three numbers in this problem:

11 yellow toy cars are on the table. 17 blue cars and 19 red cars are also on the table. How many toy cars in all are on the table?

The key to this problem was the words "in all." LC knew she had to add all the numbers to get the correct answer. But there were three numbers! LC saw she would have to add two numbers together first. Then she would take that answer and add it to the third number. LC remembered that Mr. D8N always said "look for the numbers that are easier to add." She decided to model the problem with base ten blocks on her place value mat. That way she could see the numbers more clearly:

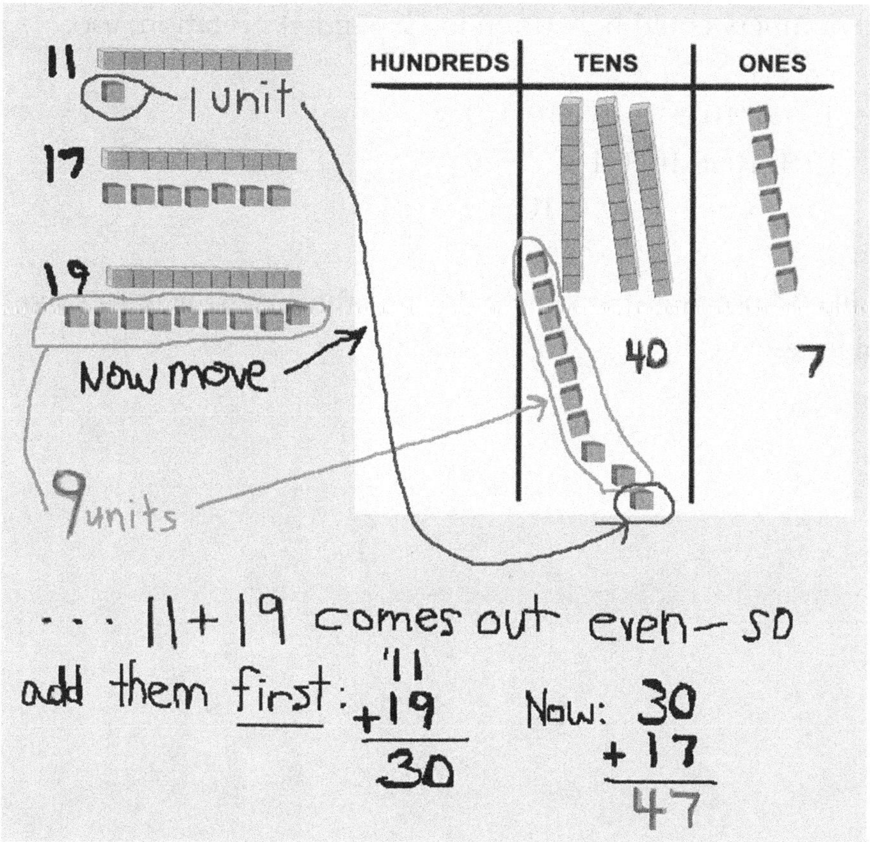

Seeing the blocks, LC knew that two of the numbers would be easiest to add first. This was because she knew to take ten unit blocks and "make ten" of them in the TENS place value column. Since there was one unit block for the number 11, and nine unit blocks for the number 19, then 1 + 9 = 10. This made four tens -- 40.

She wrote the number 40 on her mat in the TENS column. This left seven unit blocks in the ONES column. She wrote the number 7 on her mat in the ONES column. The blocks made it clear that adding the numbers in any order would give the same answer:

11 + 17 + 19 = 47 ....is the same as... 11 + 19 + 17 = 47

The numbers could be written in Expanded Notation, too:

10 + 1 + 10 + 7 + 10 + 9 =
10 + 10 + 10 + 1 + 7 + 9 =
30 + 10 + 7 = 47

LC liked knowing that she could find the easiest way to solve this problem!

Problem #7:

#7: The Planets Farm has 12 wow-trees in the front and 17 in back. How many wow-trees in all are on the Planets Farm?

The Stars Farm has 4 more wow-trees than the Planets Farm. How many wow-trees does the Stars Farm have?

25

26

Solution to Problem #7:

This problem had two parts, LC saw:

1st: **The Planets Farm has 12 wow-trees in the front and 17 in back. How many wow-trees in all are on the Planets Farm?**

2nd: **The Stars Farm has 4 more wow-trees than the Planets Farm. How many wow-trees does the Stars Farm have?**

LC decided to model both parts of the problem. She would use base ten blocks on her place value mat. First, she used the key "in all"--this told her to add the numbers 12 and 17. This would be the total sum of wow-trees on the Planet Farm. Next, LC saw that she had to compare. The number of trees on the Planet Farm was not the same as the number of trees on the Stars Farm. The problem said there were 4 <u>more</u> trees on the Stars Farm. Her work looked like this:

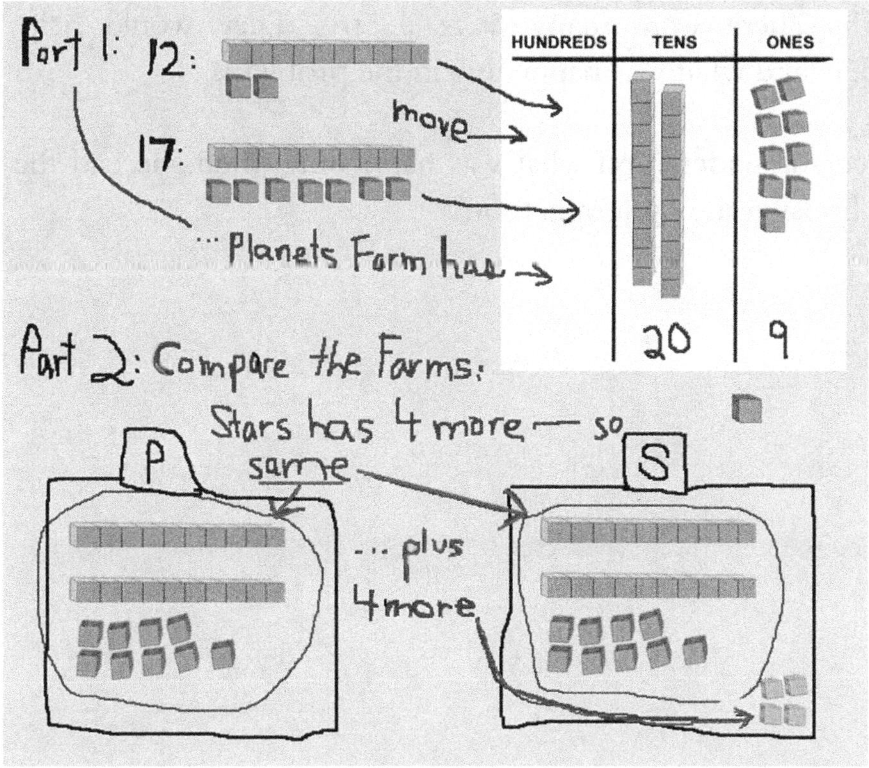

LC saw the key to the second part. She would show the same number of trees on each farm -- and then add the "4 more" that the problem said the Stars Farm had. Last, LC had to count up all the trees on the Stars Farm. The base ten blocks made that easy!

For the first part, the equation was: 12 + 17 = ___
That answer was 29. That answer was the number of trees on the Planets Farm. To get the number of trees on the Stars Farm, LC added 4. That equation was: 29 + 4 = ___

LC found that when she counted on 4 from 29, she got an answer of 33.

LC knew she could solve these problems in many ways. She could use the base ten blocks. She could use a number line. She could use counters. She could use drawings. She could use mental math. There were many ways to try. They would help LC understand what was happening in the problems.

Once she understood what was happening, then she had the key to choosing the right operation!

LC After the Test

LC took a deep breath. She had solved all of the problems!
And she had remembered to use all her ways of finding the keys.

Looking for the key words was a big help. Making a model helped, too. Those strategies gave her the operation to use: add or subtract.
LC closed up her gel cube. She smiled to herself. "I think I did well!"

27

28

29

"Well done, class!" said Mr. D8N. You can put away your gel cubes, now."

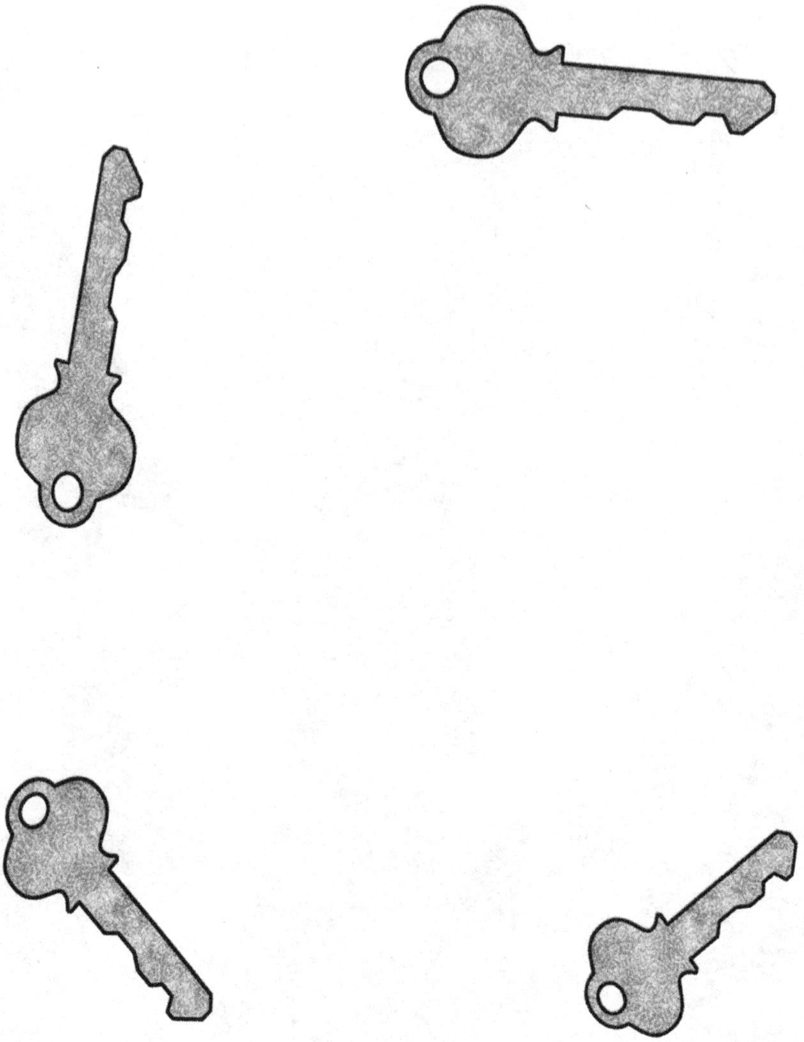

But why was Mr. D8N glowing and fading in that odd way?

30

31

A Surprise!

Elsie opened her eyes.

Oh my goodness! It had all been a dream! She wasn't a green alien. She didn't go to school in a spiral tube, but on a bus. She didn't think her answers into a gel cube--she wrote them on paper.

But one thing was still true: she did have a math test today!

32

33

Elsie jumped out of bed. She felt alert and full of energy.

Elsie could hear her mother calling her:
"Get up, Miss Elsie Foray! You have a big math test, today!"

Find the KEYS:

words that tell you what operation to use!

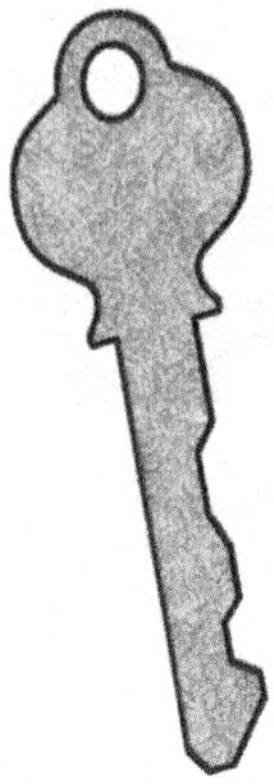

+ – x ÷

Elsie laughed out loud. She knew she could take her math test and do well. After all, she had just practiced taking that test--in her dream!

She knew what to do. She would do what her teacher Mr. Dayton always told the class to do. She would look for the key words and make models of what was happening.

Elsie smiled again. This math test was going to be fun!

***The End...*for now. Keep reading for more tips and tricks that will help you solve word problems....**

The Whiteboards in LC's Class

LC and her friends found the information on these boards to be helpful. When they had a word problem to solve, they could look at the boards for reminders about the KEYS.

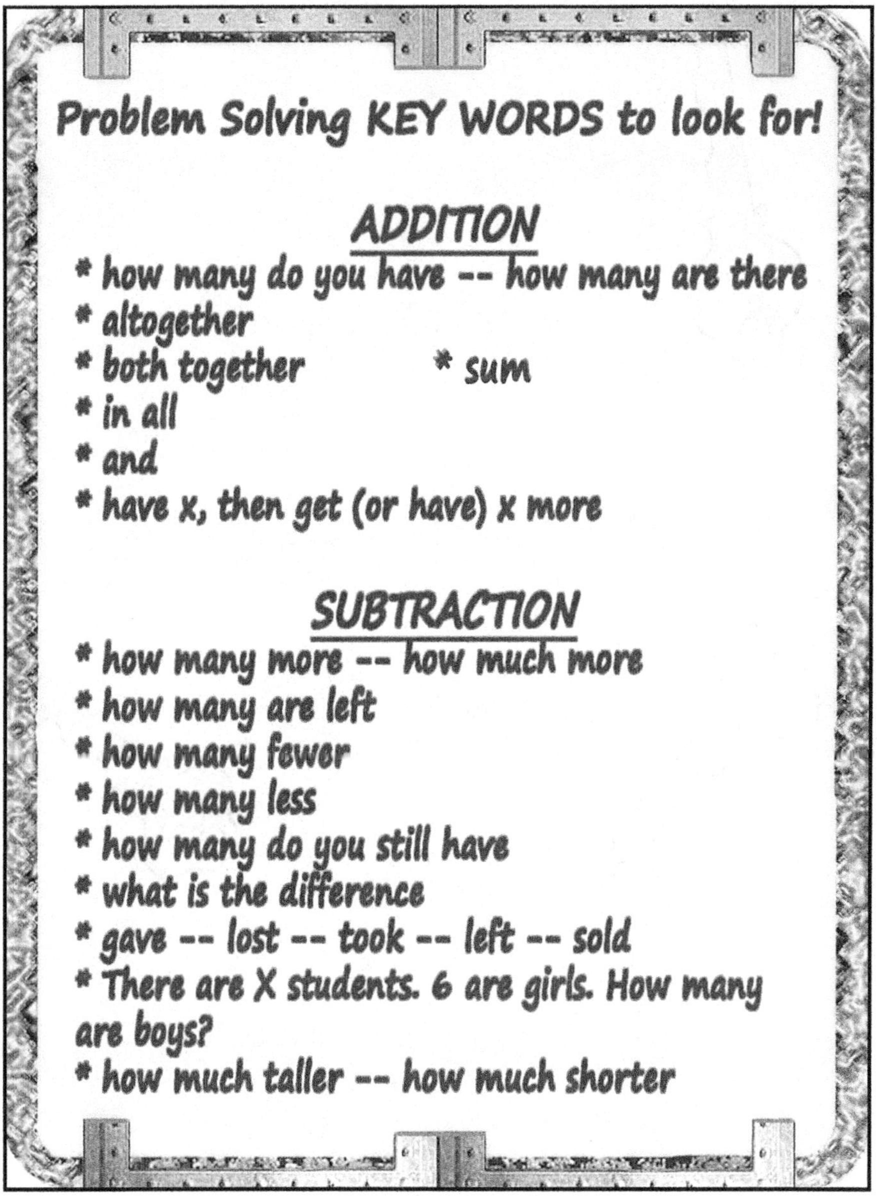

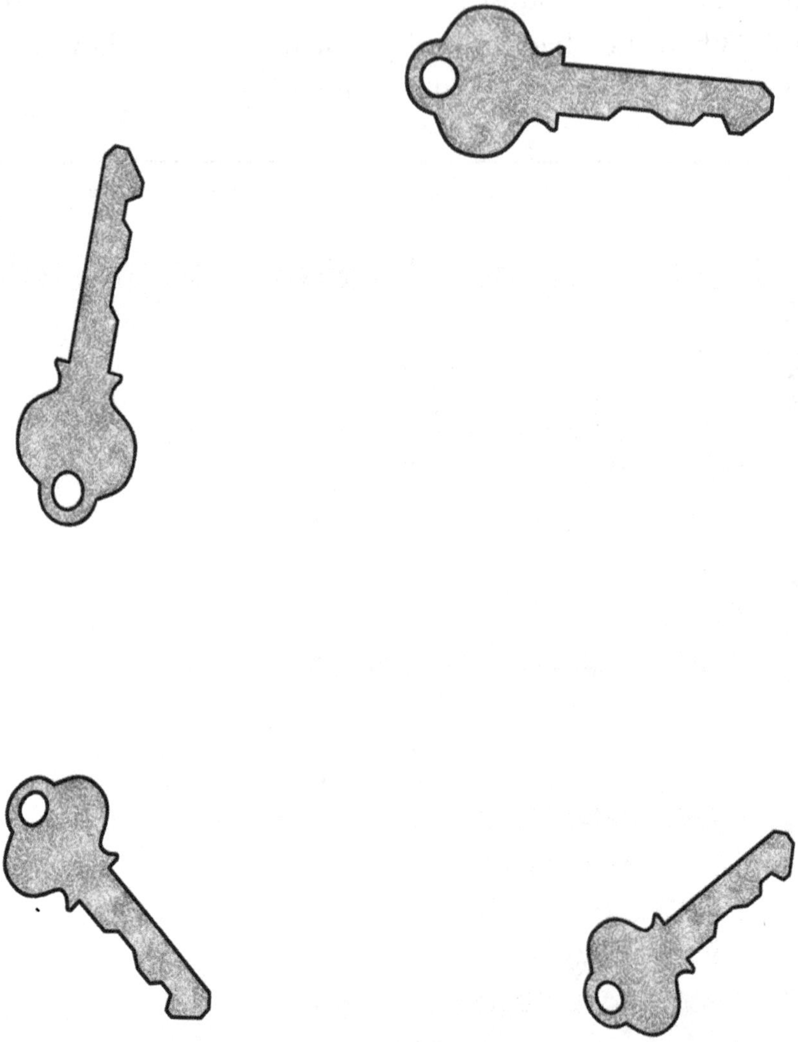

Here are two types of word problem that each use the word "more"—but that call for different operations. To solve one, you would add; to solve the other, you would subtract. The board has some good suggestions on how to get started solving these problems!

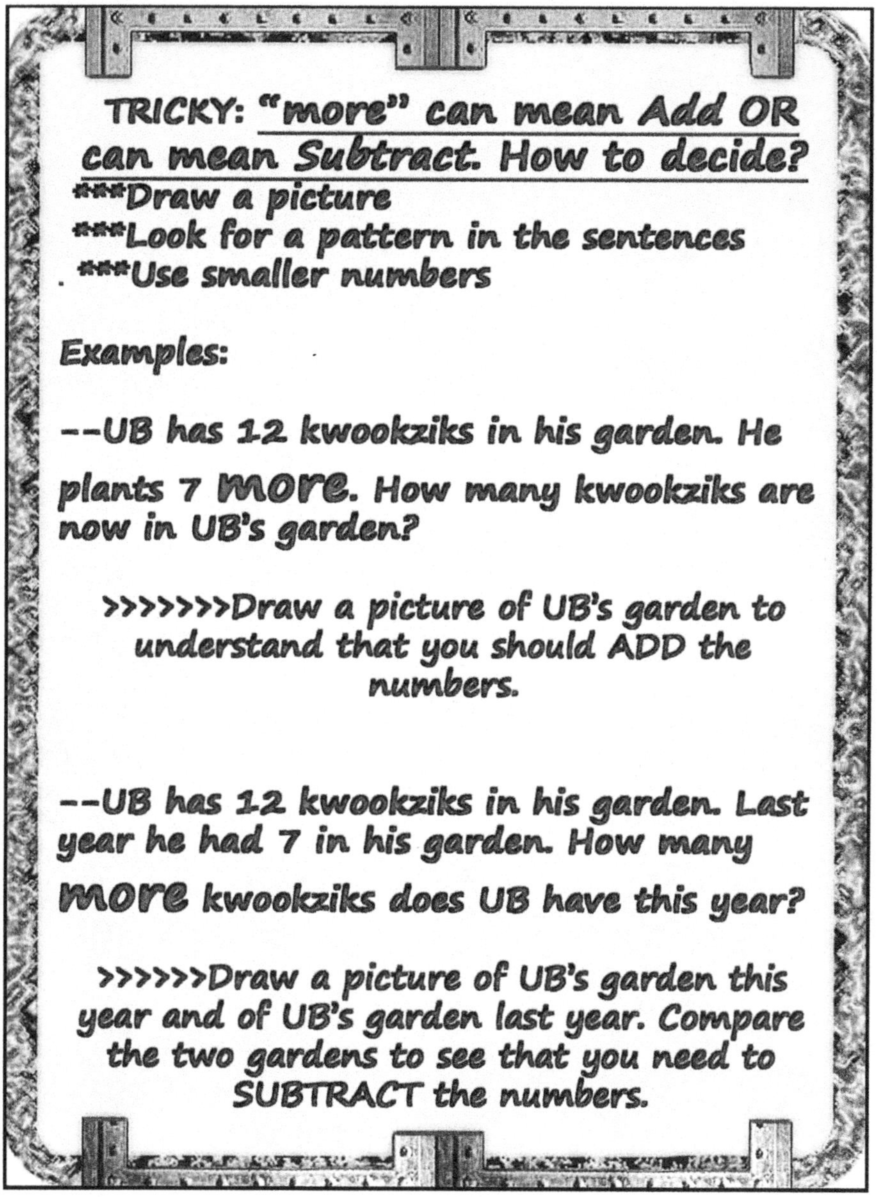

Here are two types of word problems that contain the word "more"—but that call for different operations. In [illegible] you would add [illegible] the other you would subtract [illegible] some good [illegible] how to [illegible] problems!

[illegible]

Question:

[illegible]

[illegible] more [illegible]

[illegible] SUBTRACT the numbers.

About the Mathematics: Keys to Learning

The test that LC took is centered on the second grade math curriculum, but the lessons it contains can be useful to learners working along a fairly wide range of mathematical proficiency.

Because the problems are not labeled with the operation (or operations) to be used, children thinking along with LC get practice in looking for the clues that will tell them when they need to add, and when they need to subtract.

This skill is fundamental to achieving confidence in doing mathematics: in our lives outside the schoolroom we are rarely told what we need to do in order to solve any mathematical problem that may confront us. Children need to know that they have available a number of strategies to try, including looking for key words and phrases, and modeling the problem-situation using drawings, counters, number lines, base ten blocks, and other manipulatives and techniques.

Children can develop a positive attitude toward math. A panicky feeling of not knowing where to start is the enemy of learning, though it is all too commonly accepted as "normal". Instead, children can be helped to develop the attitude "this will be fun to solve; what should I try first?"

Word problems (or story problems) are often seen as particularly challenging for children (and sometimes for adults!) When students in the early grades are developing fluency with addition, subtraction, and multiplication, they are often given pages of similar word problems.

Practice is important in learning math. But if all the problems are solved in the same way--say, by subtracting the second number mentioned in the problem from the first number mentioned--then students may work through the page on automatic pilot. What is lost is the opportunity to notice which words and phrases actually signal various crucial facts: the operation to use; the order in which operations should be applied; and which numbers should be dealt with first.

In the real world, when we need to do something with numbers, we are rarely told what operation to use or which numbers to operate on.

For example, a child may be invited to play with a group of children in which each is supposed to contribute 50 interlocking blocks to a building project. How many more blocks must the child dig out of his toy box, if he already has 14 blocks in his pockets?
The child who wants to be accepted by the others in this situation is motivated to solve for the unknown quantity--even though the operation to be used is not given to him. It doesn't appear at the top of a worksheet. It must be discovered by the child, from his understanding of the situation and the language involved. He has 14 blocks, and needs some number more to get to 50; what will he do with the numbers 14 and 50?
The solution of a real-world problem calls for not only "number sense," but "operation sense," too.

For some children the solution to this problem will seem instinctive. The child has 14; 14 plus some unknown number will result in 50--the target number of blocks. Academically, this can be referred to as a "Join - Change Unknown" problem. The classification itself doesn't matter; what is useful about the classifications is that they remind us to give children opportunities to confront many types of problems. In the real world, we have to decide what operation to use and what numbers to use in contexts that vary. Here are some examples of those variations, with the classifications often applied to them:

JOIN - RESULT UNKNOWN: Three cats sat on a rug. Two more walked over and sat on the rug. How many cats are on the rug now? ... 3 + 2 = ___
SEPARATE - RESULT UNKNOWN: Five cats sat on a rug. Three walked away. How many cats are on the rug now? ..... 5 - 3 = ___
JOIN - CHANGE UNKNOWN: Three cats sat on a rug. Some more cats walked over and joined them. Then there were five cats on the rug. How many cats walked over and joined? ..... 3 + ___ = 5
SEPARATE - CHANGE UNKNOWN: Five cats sat on a rug. Some left. Then there were two cats. How many left? ..... 5 - ___ = 2
JOIN - START UNKNOWN: Some cats sat on a rug. Two more walked over and joined them. Then there were five cats on the rug. How many cats were on the rug before? ..... ___ + 2 = 5
SEPARATE - START UNKNOWN: Some cats sat on a rug. Three left. Then there were two cats on the rug. How many cats were on the rug to begin with? ..... ___ - 3 = 2

(In each of the problems in this section, there is one "type of thing" that changes in number over time.)

PART-PART-WHOLE - TOTAL UNKNOWN: Three red marbles and two blue marbles sit on the table. How many marbles are on the table? ..... 3 + 2 = ___

PART-PART-WHOLE - ADDEND UNKNOWN: Five marbles are on the table. Three are red and the rest are blue. How many marbles are blue? ..... 3 + ___ = 5 and 5 - 3 = ___

PART-PART-WHOLE - BOTH ADDENDS UNKNOWN: Delora has five marbles. How many can she put in her left pocket and how many in her right pocket? ..... 5 = 0 + 5; 5 = 5 + 0; 5 = 1 + 4; 5 = 4 + 1; 5 = 2 + 3; 5 = 3 + 2

(In each of the problems in this section, there are two "types of thing" at the same time.)

COMPARE - DIFFERENCE UNKNOWN: "more" version: Sam has two books. Riva has five books. How many more books does Riva have than Sam? ...."fewer" version: Sam has two books. Riva has five books. How many fewer books does Sam have than Riva? ..... 2 + ___ = 5; 5 - 2 = ___

COMPARE - BIGGER UNKNOWN: "more" version: Riva has three more books than Sam. Sam has two books. How many books does Riva have? ...."fewer" version: Sam has three fewer books than Riva. Sam has two books. How many books does Riva have? ..... 2 + 3 = ___; 3 + 2 = ___

COMPARE - SMALLER UNKNOWN: "more" version: Riva has three more books than Sam. Riva has five books. How many books does Sam have? .... Sam has three fewer books than Riva. Riva has five books. How many books does Sam have? ..... 5 - 3 = ___; ___ + 3 = 5

(In each of the problems in this section, there are two people who each, at the same time, have differing numbers of the same "type of thing.")

<u>It is **not** at all important for children (or even parents) to know the names of these classifications.</u> The value of the classifications lies, instead, in reminding us that children need practice in ALL these varied ways of dealing with similar situations. Each represents a differing type of real-life situation: sometimes we're looking at a change over time, while other times we're looking at a comparison at the same time. Sometimes we know the number we're starting with, while other times we know the number we end with.

Sometimes we're looking at combinations of numbers that make up a larger number, while other times we're looking at the result of the actions of joining or of taking away.

Practice in all the variations help children gain confidence in their ability to deal with problems.

The problems in this book have the following classifications:
#1: The eye-fish -- PART-PART-WHOLE, ADDEND UNKNOWN
#2: The pwerts -- SEPARATE, CHANGE UNKNOWN
#3: The fly-boards -- JOIN, START UNKNOWN
#4: The treats -- COMPARE, DIFFERENCE UNKNOWN
#5: The zilk-zorks -- COMPARE, SMALLER UNKNOWN
#6: The toy cars -- PART-PART-WHOLE, TOTAL UNKNOWN (with 3 types; larger numbers)
#7: The wow-trees -- Step 1: JOIN, RESULT UNKNOWN .....Step 2: COMPARE, BIGGER UNKNOWN

As LC confronted these problems, she showed the value of trying to understand what was happening in the problem. Models of the problem situation (made with counters, crayons, base ten blocks, and other materials) helped LC feel confident that she was applying the correct operation to the numbers. Attention to the key words and phrases was important to LC in making her decisions about what to do first, and in finding the problem solutions.

When children start to see math problems as interesting puzzles to solve, and as chances to employ a strategy from a tool kit of strategies they've mastered, then they are well on the way to success with mathematics.

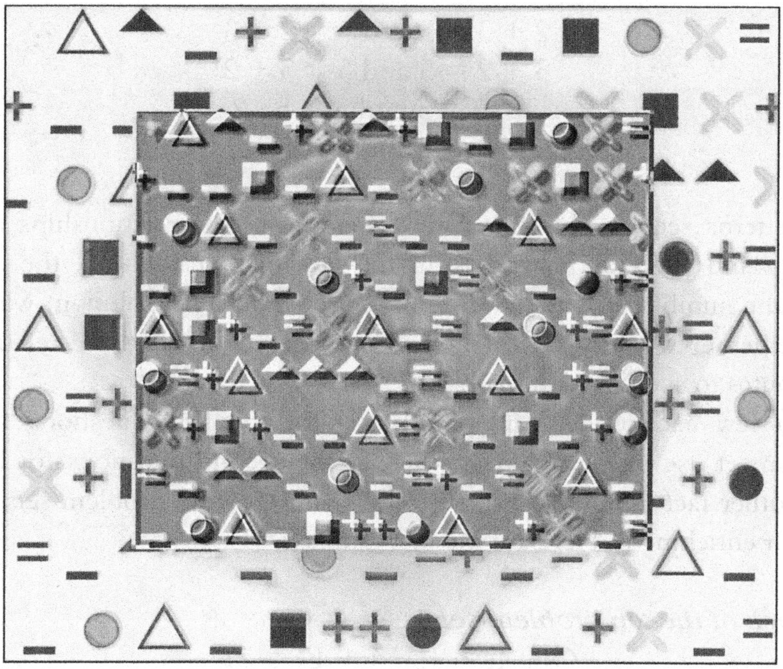

PROBLEM SETS

As discussed in the section "About the Mathematics: Keys to Learning," it's good for learners to get practice in deciding what operation to use, and what numbers to work with first. An ideal type of practice is found in problems sets that use the same basic situations and objects—but that require differing choices of operation and order.

These ten sets of problems provide that kind of practice. Each set contains twelve questions: one each of the twelve types listed in "About the Mathematics." For each set, the order of types is different. The goal is for learners to develop confidence about their ability to find key words and contexts; these will let learners decide what operation to use. Solving by thinking through the key words and other clues, as opposed to solving in a rote fashion, builds confidence.

FACT FAMILIES

For addition and subtraction, three differing numbers create a "family" of related math sentences (or *equations*). To illustrate, the fact family sentences for 2, 3, and 5 are:

$$2 + 3 = 5 \text{ and } 5 = 2 + 3$$
$$3 + 2 = 5 \text{ and } 5 = 3 + 2$$
$$5 - 2 = 3 \text{ and } 3 = 5 - 2$$
$$5 - 3 = 2 \text{ and } 2 = 5 - 3$$

The patterns seen in this fact family stem from the relationships between addition and subtraction, and from the fact that in addition, the order in which the numbers are added doesn't matter. The sum (solution) will be the same whether we add 2 + 3 or 3 + 2. Recognition of these relationships contributes to success with math.

With a few exceptions (for space), each of the first questions for each problem set use one fact family; each of the second questions for each set use another fact family, and so on. This feature of the problem sets may be used for enrichment. Here are the fact families:

For each of the ten problem sets:

Question 1: 4 + 17 = 21
Question 2: 3 + 5 = 8
Question 3: 2 + 9 = 11
Question 4: 5 + 13 = 18
Question 5: 10 + 19 = 29
Question 6: 14 + 21 = 35
Question 7: 4 + 8 = 12
Question 8: 3 + 6 = 9
Question 9: 4 + 12 = 16
Question 10: 11 + 15 = 26
Question 11: 7 + 8 = 15
Question 12: 5 + 22 = 27

This feature means that learners will see ten problems for each fact family. Possible enrichment activities might include making up new problems for a fact family, or re-writing any single existing problem (of the learner's choice) using each of the twelve provided fact families. What's important: to keep the focus on finding key words, and on modeling the situation (with drawings, counters, number lines, etc.).

For each problem, learners should write an open sentence that "translates" the English of the word problem into math symbols. For example, for the problem "Five cats sat on a rug, then some left. Now there are two cats. How many cats left?" the corresponding open sentence is "5 - __ = 2".

(Note that it would not be incorrect to write "2 = 5 - __" or even "5 – 2 = __" instead, because all three equations are mathematically equivalent. On the Solutions page for each problem set, more than one correct open sentence "translation" is shown for many of the problems.)

After writing an open sentence which translates the word problem into math symbols, the learner should solve for the unknown number.

Modeling a word problem can be a very helpful step in the process of finding a solution, as LC4A's experience shows. For drawings and number lines, all that is needed is paper and pencil or crayon. For modeling with counters and base-ten blocks, options range from store-bought "math manipulatives" to online "virtual manipulatives" to homemade alternatives: coins or buttons as counters, and paper representations of base-ten blocks.

MATH MANIPULATIVES

Many commercial outlets sell counters, base-ten blocks, pattern blocks, fraction circles, Cuisenaire rods, snap cubes, algebra tiles, and other math manipulatives.

In addition, many websites offer "virtual manipulatives" with activities. Some of these sites charge a fee and some are free. Access to a computing device is required for using these, of course. One frequently-recommended site is the National Library of Virtual Manipulatives, which is free of charge, and can be found at http://nlvm.usu.edu/en/nav/vlibrary.html

A free—and also low-tech—alternative is to make paper manipulatives. Many sites offer printable pages which can be cut into (for example) sets of base-ten blocks: units (representing "1" each), rods ("10"), and flats ("100"). Possibly the most frequently-linked source for such printable pages is found at http://lrt.ednet.ns.ca/PD/BLM/table_of_contents.htm and is sponsored by the Learning Resources and Technology Services of the Nova Scotia Department of Education. This page contains well over five hundred links to printable pages offering not only base-ten blocks and place-value mats, but tangrams, pattern blocks, number lines, hundreds charts, geoboards, fraction circles, clocks, flash cards, calendars, ten-frames, multiplication tables, and other useful printable items such as graph paper and dot paper.

Other sources of free printable math-related supplies can be found by searching "math manipulatives" and "printable."

SOME FINAL ADVICE AND ENCOURAGEMENT:

A great advantage for learners who form the habit of modeling word problems is the self-assurance that comes from knowing where to begin. The process starts with reading the problem over; reading it several times can be a good idea. The key words and phrases will become clear with re-reading; students will learn to look for "more than," "less than," "joined," "left," "how many in all," "how many more," and other clues to what is happening in the word problem. (See the Whiteboards pages of this book for useful lists and tips on key words.) Making a model of what is taking place in the problem can be a big help. In addition and subtraction problems, there are usually three basic variations on what is happening:

- There is one type of thing, and it changes in number over time. (Example: cats on a rug, joining or leaving.)
- There are two types of things at the same time. (Example: Red marbles and blue marbles on a table.)
- There are two people who, at the same time, have different numbers of the same type of thing. (Example: Sam and Riva each have some books.)

Each of these situations can be modeled. Making a model can help make it clear what to do first.

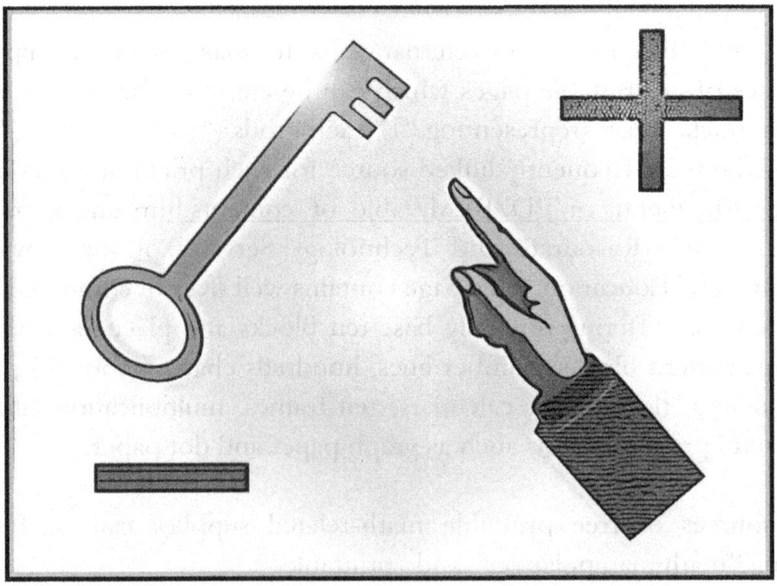

SET 1: For each problem, write an open sentence (example: 3 + __ = 5) and solve. Look for key words. You can model the problem in many ways: draw a picture; use counters or base-ten blocks; make a number line, etc.

1. A bowl holds 21 grapes. Mai takes 4 grapes from the bowl. How many grapes are in the bowl now?

2. 8 grapes are in a bowl. Dan takes some out. Now there are 5 grapes in the bowl. How many grapes did Dan take?

3. Dan has 9 grapes. Mai has 11 grapes. How many more grapes does Mai have than Dan has?

4. Mai has 5 grapes on her plate. How many can she hold in her left hand and how many in her right hand? Name all the possibilities. (For example: 0 grapes in her left hand and 5 grapes in her right hand: 5 = 0 + 5.)

5. 19 grapes are in a bowl. Then 10 more grapes are added to the bowl. How many grapes are in the bowl now?

6. There are 21 fewer grapes on the blue plate than there are on the yellow plate. There are 35 grapes on the yellow plate. How many grapes are there on the blue plate? (Note: fewer = less)

7. There are 8 grapes on a plate. Then Mai puts more grapes on the plate. Now there are 12 grapes on the plate. How many did Mai put on the plate?

8. Dora has 3 more grapes than Ali. Ali has 6 grapes. How many grapes does Dora have?

9. 16 grapes are in a bowl. 12 are green and the rest are red. How many of the grapes are red?

10. Some grapes sit in a bowl. Dan put 15 more grapes into the bowl. Then there were 26 grapes in the bowl. How many were in the bowl to start?

11. One plate on a table holds 7 grapes, and another plate on the table holds 8 grapes. How many grapes are on the table?

12. Some grapes are in a bowl. Dan removed 5 grapes from the bowl. Then there were 22 grapes in the bowl. How many grapes were in the bowl to begin with?

Workspace; and a NOTE: Solutions for the GRAPES problems will be visible when you turn the page.

This is true for all the problem sets: the solutions immediately follow the problems, to prevent the annoyance of having to flip back and forth from the problems to the back-of-the-book.

Remember that the value of working out these problems is greater than simply coming up with a number that answers the problem-question—important as that is. Accidentally glimpsing the numerical answer doesn't take away from the learning, because there's more to becoming a word-problem solver than just writing down a number.

It will help you to develop your skills if you model the problems and come up with equations that show what is happening in the problem situation. This will give you practice in spotting key words and phrases. That's what word problem experts do!

SOLUTIONS and Problem Types for SET 1: Grapes

Set 1	OPEN SENTENCES	Solution	Classification (**<u>not</u>** important for students to learn); listed to show variety
1	21 – 4 = __	17	SEPARATE – RESULT UNKNOWN
2	8 - __ = 5 or 8 - 5 = __	3	SEPARATE – CHANGE UNKNOWN
3	11 - 9 = ___ or 9 + __ = 11	2	COMPARE: DIFFERENCE UNKNOWN
4	5 = 0 + 5 = 5 + 0 5 = 1 + 4 = 4 + 1 5 = 2 + 3 = 3 + 2	See left	PART-PART-WHOLE: Both ADDENDS UNKNOWN
5	19 + 10 = __ or 10 + 19 = __	29	JOIN – RESULT UNKNOWN
6	35 – 21 = __ or 21 + __ = 35	14	COMPARE: SMALLER UNKNOWN
7	8 + __ = 12 or 12 – 8 = __	4	JOIN – CHANGE UNKNOWN
8	6 + 3 = __ or 3 + 6 = __	9	COMPARE: BIGGER UNKNOWN
9	16 = 12 + __ or 16 – 12 = __	4	PART-PART-WHOLE: ADDEND UNKNOWN
10	__ + 15 = 26 or 26 – 15 = __	11	JOIN – START UNKNOWN
11	7 + 8 = __	15	PART-PART-WHOLE: TOTAL UNKNOWN
12	__ - 5 = 22 or 22 + 5 = __	27	SEPARATE – START UNKNOWN

A Bus Can Take
You There!
Ride the
BUS!

SET 2: For each problem, write an open sentence (example: 3 + __ = 5) and solve. Look for key words. You can model the problem in many ways: draw a picture; use counters or base-ten blocks; make a number line, etc.

1. Some busses are in a bus station. 17 busses are driven away. 4 busses remain. How many busses were in the station to begin with?

2. There are 5 busses in the left side of the Lake bus station and 3 busses in the right side of the Lake bus station. How many busses are in the Lake bus station altogether?

3. There are 2 busses in the Fairgrounds bus station. Then more busses are driven into the bus station. Now there are 11 busses in the Fairgrounds station. How many busses were driven in?

4. The North bus station has 13 more busses than the South bus station has. The South bus station has 5 busses. How many busses are in the North bus station?

5. 29 busses are in the East bus station. 10 of the busses are new and the rest of the busses are old. How many of the busses in the East bus station are old?

6. 35 busses are in the West bus station. Some of those busses are driven out of the station. Now there are 14 busses in the station. How many were driven away?

7. Some busses are in the Central bus station. Then 4 more busses were driven into the station. Now there are 12 busses in the station. How many busses were there in Central bus station to start?

8. There are 9 busses in the Hilltop bus station. There are 6 fewer busses in the Valley bus station than there are in the Hilltop bus station. How many busses are in the Valley bus station?

9. There are 16 busses in the College bus station. Now 12 busses are driven away. How many busses are in the College bus station now?

10. 11 busses sit in the Park bus station. Now 15 more busses are driven into the station. How many busses are in the Park bus station now?

11. 7 bus drivers want to park their busses in two rows: Row A and Row B. How many ways can they arrange the 7 busses in the two rows? List all the possible ways. (For example: 0 busses in Row A and 7 busses in Row B: $7 = 0 + 7$.)

12. There are 5 busses parked in Downtown bus station and 27 busses parked in Uptown bus station. How many more busses are parked in Uptown bus station than in Downtown bus station?

SOLUTIONS and Problem Types for SET 2: Busses (more than one bus)

Set 2	OPEN SENTENCES	Solution	Classification (**not** important for students to learn); listed to show variety of problems.
1	___ - 17 = 4 or 17 + 4 = ___	21	SEPARATE – START UNKNOWN
2	5 + 3 = ___	8	PART-PART-WHOLE: TOTAL UNKNOWN
3	2 + ___ = 11 or 11 – 2 = ___	9	JOIN – CHANGE UNKNOWN
4	5 + 13 = ___ or 13 + 5 = ___	18	COMPARE: BIGGER UNKNOWN
5	29 = 10 + ___ or 29 – 10 = ___	19	PART-PART-WHOLE: ADDEND UNKNOWN
6	35 - ___ = 14 or 35 – 14 = ___	21	SEPARATE – CHANGE UNKNOWN
7	___ + 4 = 12 or 12 – 4 = ___	8	JOIN – START UNKNOWN
8	9 – 6 = ___ or 6 + ___ = 9	3	COMPARE: SMALLER UNKNOWN
9	16 – 12 = ___	4	SEPARATE – RESULT UNKNOWN
10	11 + 15 = ___ or 15 + 11 = ___	26	JOIN – RESULT UNKNOWN
11	7 = 0 + 7 = 7 + 0 7 = 1 + 6 = 6 + 1 7 = 2 + 5 = 5 + 2 7 = 3 + 4 = 4 + 3	See left	PART-PART-WHOLE: Both ADDENDS UNKNOWN
12	27 – 5 = ___ or 5 + ___ = 27	22	COMPARE: DIFFERENCE UNKNOWN

SET 3: For each problem, write an open sentence (example: 3 + __ = 5) and solve. Look for key words. You can model the problem in many ways: draw a picture; use counters or base-ten blocks; make a number line, etc.

1. In the deep part of the pond there are 4 frogs. In the shallow part of the pond there are 17 frogs. How many frogs are there altogether?

2. There are 8 frogs sitting on a log. There are 5 fewer frogs sitting on a rock than frogs sitting on the log. How many frogs are on the rock?

3. 2 frogs are in a pond. Then 9 more frogs hop into the pond. How many frogs are in the pond now?

4. Some frogs are in a pond. Then 13 frogs hop out of the pond. Now there are 5 frogs in the pond. How many frogs were in the pond to start?

5. Some frogs are in a pond. Then 10 more frogs hop into the pond. Now there are 29 frogs in the pond. How many frogs were in the pond to start?

6. There are 14 frogs on a log. The log has two areas: sunny, and shady. The frogs move along the log so that they will be in sun or in shade. What are all the possibilities for frogs on the log? (Example: 0 frogs in the sun and 14 frogs in the shade: 14 = 0 + 14. 1 frog in the sun and 13 frogs in the shade: 14 = 1 + 13. List all the ways the frogs can sit.)

7. The old pond has 12 frogs in it and the new pond has 4 frogs in it. How many more frogs are in the old pond than in the new pond?

8. 9 frogs are sitting on a rock. Then 3 frogs hop off the rock. How many frogs are on the rock now?

9. 16 frogs are swimming in a pond. Then some frogs leave the pond. Now there are 4 frogs still swimming in the pond. How many frogs left?

10. There are 26 frogs in a pond. 15 of the frogs are green and the rest are brown. How many of the frogs are brown?

11. In a pond, frogs sit on a log and on a rock. 7 more frogs sit on the log than on the rock. There are 8 frogs sitting on the rock. How many frogs sit on the log?

12. There are 22 frogs sitting in a pond. Then some more frogs hop into the pond. Now there are 27 frogs sitting in the pond. How many frogs hopped in?

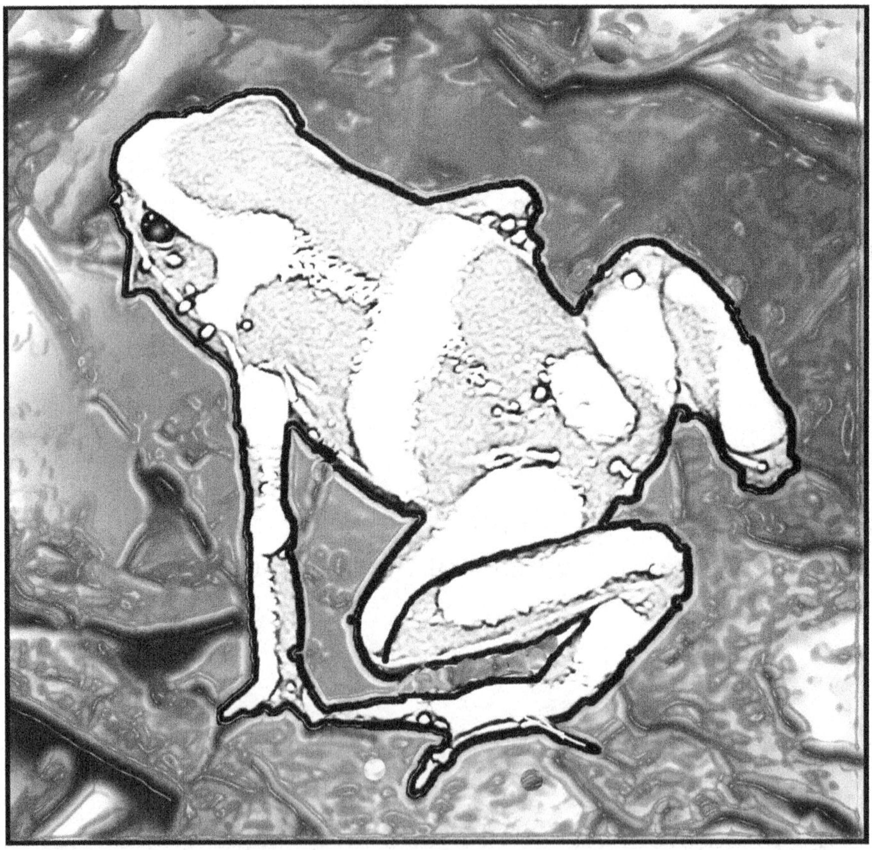

SOLUTIONS and Problem Types for SET 3: Frogs

Set 3	OPEN SENTENCES	Solution	Classification (**not** important for students to learn); listed to show variety of problems.
1	4 + 17 = __	21	PART-PART-WHOLE: TOTAL UNKNOWN
2	8 – 5 = __ or 5 + __ = 8	3	COMPARE: SMALLER UNKNOWN
3	2 + 9 = __ or 9 + 2 = __	11	JOIN – RESULT UNKNOWN
4	__ - 13 = 5 or 5 + 13 = __	18	SEPARATE – START UNKNOWN
5	__ + 10 = 29 or 29 – 10 = __	19	JOIN – START UNKNOWN
6	14 = 0 + 14 = 14 + 0 14 = 1 + 13 = 13 + 1 14 = 2 + 12 = 12 + 2 14 = 3 + 11 = 11 + 3 14 = 4 + 10 = 10 + 4 14 = 5 + 9 = 9 + 5 14 = 6 + 8 = 8 + 6 14 = 7 + 7	See left	PART-PART-WHOLE: Both ADDENDS UNKNOWN
7	12 – 4 = __ or 4 + __ = 12	8	COMPARE: DIFFERENCE UNKNOWN
8	9 – 3 = __	6	SEPARATE – RESULT UNKNOWN
9	16 - __ = 4 or 16 – 4 = __	12	SEPARATE – CHANGE UNKNOWN
10	26 = 15 + __ or 26 – 15 = __	11	PART-PART-WHOLE: ADDEND UNKNOWN
11	8 + 7 = __ or 7 + 8 = __	15	COMPARE: BIGGER UNKNOWN
12	22 + __ = 27 or 27 – 22 = __	5	JOIN – CHANGE UNKNOWN

SET 4: For each problem, write an open sentence (example: 3 + __ = 5) and solve. Look for key words. You can model the problem in many ways: draw a picture; use counters or base-ten blocks; make a number line, etc.

1. Some mugs sit on a shelf. Grace puts 4 mugs on the shelf. Now there are 21 mugs on the shelf. How many were on the shelf to begin with?

2. There are 5 mugs on the table. Then Jamal puts more mugs on the table. Now there are 8 mugs on the table. How many mugs did Jamal put there?

3. The top shelf has 2 more mugs than the bottom shelf. On the bottom shelf are 9 mugs. How many mugs are on the top shelf?

4. 18 mugs are on the shelf. 5 of the mugs are large, and the rest are small. How many of the mugs are small?

5. Some mugs are in a store. Jamal buys 10 of them to give as presents. Now there are 19 mugs left in the store. How many mugs were in the store to start?

6. 21 mugs sit on the left side of a shelf, and 14 mugs sit on the right side of the shelf. How many mugs are on the shelf altogether?

7. There are 12 mugs on the top shelf. There are 8 fewer mugs on the bottom shelf than there are on the top shelf. How many mugs are there on the bottom shelf?

8. On the table are 9 mugs. Grace takes away 6 of the mugs. How many mugs are on the table now?

9. 16 mugs are for sale in a store. Some of the mugs are sold. Now there are 12 mugs left for sale in the store. How many mugs were sold?

10. The bottom shelf holds 15 mugs. The top shelf holds 26 mugs. How many more mugs does the top shelf hold?

11. 7 mugs sit on a shelf. Then Jamal puts 8 more mugs on the shelf. How many mugs are on the shelf now?

12. Grace has 5 mugs. She wants to arrange her mugs on the top shelf and on the bottom shelf. What are all the ways she can arrange her mugs? (Example: 0 mugs on the top shelf and 5 on the bottom shelf: 5 = 0 + 5; 1 mug on the top shelf and 4 mugs on the bottom shelf: 5 = 1 + 4. List all the ways the mugs can be put on the two shelves.)

SOLUTIONS and Problem Types for SET 4: Mugs

Set 4	OPEN SENTENCES	Solution	Classification (**not** important for students to learn); listed to show variety of problems.
1	__ + 4 = 21 or 21 – 4 = __	17	JOIN – START UNKNOWN
2	5 + __ = 8 or 8 – 5 = __	3	JOIN – CHANGE UNKNOWN
3	9 + 2 = __	11	COMPARE: BIGGER UNKNOWN
4	18 = 5 + __ or 18 – 5 = __	13	PART-PART-WHOLE: ADDEND UNKNOWN
5	__ - 10 = 19 or 19 + 10 = __	29	SEPARATE – START UNKNOWN
6	21 + 14 = __	35	PART-PART-WHOLE: TOTAL UNKNOWN
7	12 – 8 = __ or 8 + __ = 12	4	COMPARE: SMALLER UNKNOWN
8	9 – 6 = __	3	SEPARATE – RESULT UNKNOWN
9	16 - __ = 12 or 16 – 12 = __	4	SEPARATE – CHANGE UNKNOWN
10	26 – 15 = __ or 15 + __ = 26	11	COMPARE: DIFFERENCE UNKNOWN
11	7 + 8 = __ or 8 + 7 = __	15	JOIN – RESULT UNKNOWN
12	5 = 0 + 5 = 5 + 0 5 = 1 + 4 = 4 + 1 5 = 2 + 3 = 3 + 2	See left	PART-PART-WHOLE: Both ADDENDS UNKNOWN

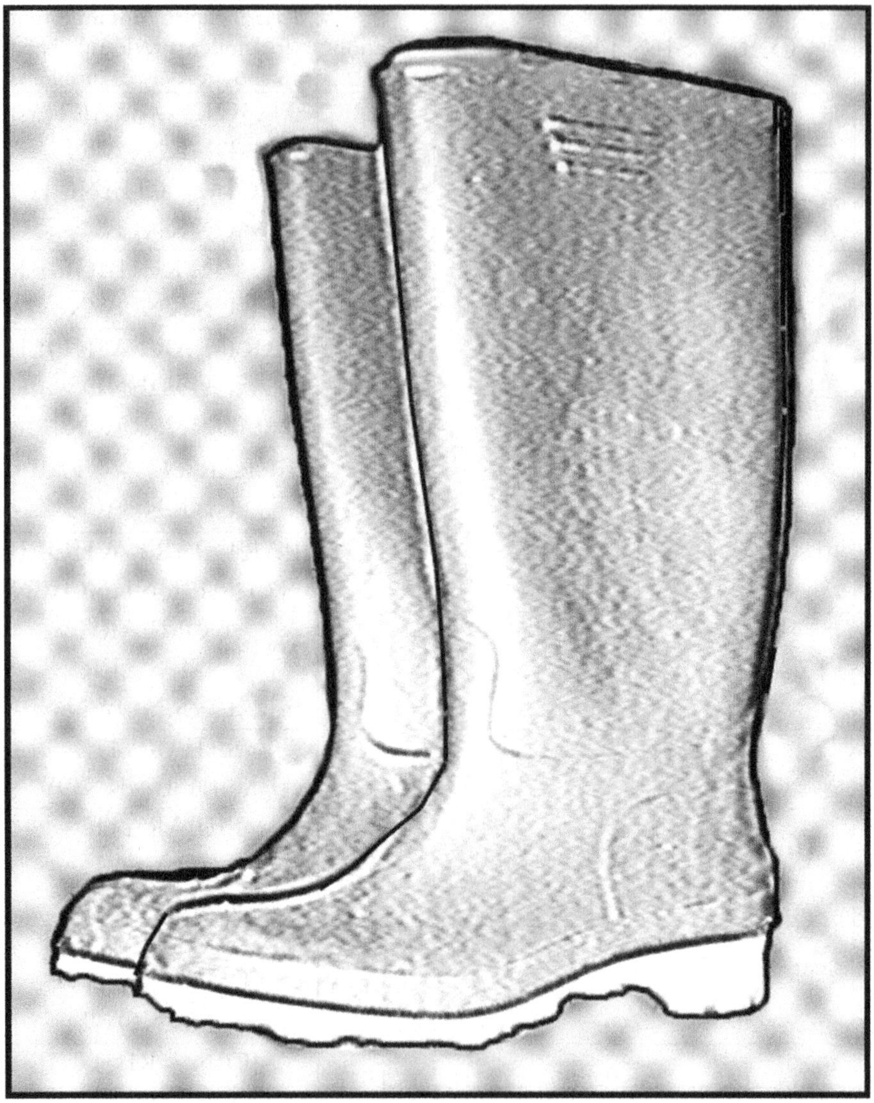

SET 5: For each problem, write an open sentence (example: 3 + __ = 5) and solve. Look for key words. You can model the problem in many ways: draw a picture; use counters or base-ten blocks; make a number line, etc.

1. 21 pairs of boots are for sale in a store. 4 of the pairs of boots are short and the rest are tall. How many pairs of boots are tall?

2. Ken has 5 more pairs of boots than Sina has. Sina has 3 pairs of boots. How many pairs of boots does Ken have?

3. There are 9 pairs of boots in the store window. Then a worker puts more boots in the window. Now there are 11 pairs of boots in the store window. How many pairs did the worker put there?

4. 5 pairs of boots are on a shelf. Then 13 more pairs of boots are put on the shelf. How many pairs of boots are on the shelf now?

5. There are 29 pairs of boots in the store window of Shoes-4-You. There are 19 fewer pairs of boots than that, in the store window of K-Shoes. How many pairs of boots are there in the store window of K-Shoes?

6. 35 pairs of boots are in the store window of Bootsy-Boots. A worker takes out some of the pairs of boots. Now there are 21 pairs of boots left in the store window. How many pairs of boots did the worker take out?

7. Lana wants to put her 4 pairs of boots in her closet. She can put boots on the left side or on the right side. What are all the ways she can put the pairs of boots in the closet? (Example: she can put 0 pairs on the left and 4 pairs on the right: $4 = 0 + 4$; and so on.)

8. Dora has 9 pairs of boots. Ali has 3 pairs of boots. How many more pairs of boots does Dora have?

9. A store has 4 pairs of short boots for sale. The store has 12 pairs of tall boots for sale. How many pairs of boots are for sale in the store?

10. Some pairs of boots are in a store. Then 11 pairs of boots are sold. Now there are 15 pairs of boots still left in the store. How many pairs of boots were there to begin with?

11. Some pairs of boots are in a store window. Then a worker puts 7 more pairs of boots into the window. Now there are 15 pairs of boots in the window. How many pairs of boots were in the window before?

12. A store has 27 pairs of boots for sale. Then 5 pairs are sold. How many pairs of boots are there in the store now?

SOLUTIONS and Problem Types for SET 5: Boots

Set 5	OPEN SENTENCES	Solution	Classification (**not** important for students to learn); listed to show variety of problems.
1	21 = 4 + __ or 21 – 4 = __	17	PART-PART-WHOLE: ADDEND UNKNOWN
2	3 + 5 = __	8	COMPARE: BIGGER UNKNOWN
3	9 + __ = 11 or 11 – 9 = __	2	JOIN – CHANGE UNKNOWN
4	5 + 13 = __ or 13 + 5 = __	18	JOIN – RESULT UNKNOWN
5	29 – 19 = __ or 19 + __ = 29	10	COMPARE: SMALLER UNKNOWN
6	35 - __ = 21 or 35 – 21 = __	14	SEPARATE – CHANGE UNKNOWN
7	4 = 0 + 4 = 4 + 0 4 = 1 + 3 = 3 + 1 4 = 2 + 2	See left	PART-PART-WHOLE: Both ADDENDS UNKNOWN
8	9 – 3 = __ or 3 + __ = 9	6	COMPARE: DIFFERENCE UNKNOWN
9	4 + 12 = __	16	PART-PART-WHOLE: TOTAL UNKNOWN
10	__ - 11 = 15 or 15 + 11 = __	26	SEPARATE – START UNKNOWN
11	__ + 7 = 15 or 15 – 7 = __	8	JOIN – START UNKNOWN
12	27 – 5 = __	22	SEPARATE – RESULT UNKNOWN

SET 6: For each problem, write an open sentence (example: 3 + __ = 5) and solve. Look for key words. You can model the problem in many ways: draw a picture; use counters or base-ten blocks; make a number line, etc.

1. Lee has 4 more blocks than Iris has. Iris has 17 blocks. How many does Lee have?

2. 5 blocks are on the table. Then 3 more blocks are put on the table. How many blocks are on the table now?

3. 2 blue blocks and 9 orange blocks are in a box. How many blocks in all are in the box?

4. There are 13 fewer yellow blocks on the table than there are green blocks. There are 18 green blocks on the table. How many yellow blocks are on the table?

5. There are 19 blocks on the floor. Then Lee puts some more blocks on the floor. Now there are 29 blocks on the floor. How many blocks did Lee put there?

6. Some blocks were in a box. Then Iris took 21 blocks out of the box. Now there are 14 blocks in the box. How many were there before?

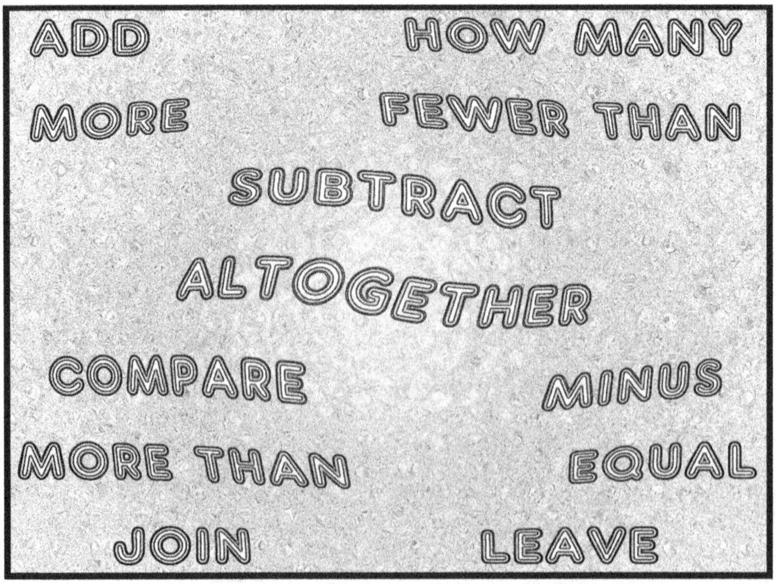

7. Some blocks were on the table. Then Lee put 8 more blocks on the table. Now there are 12 blocks on the table. How many were there to begin with?

8. Iris has 6 blocks. She wants to put them in her pockets. List all the ways that Iris can put the blocks in her pockets. (Example: 0 blocks in her left pocket and 6 blocks in her right pocket: 6 = 0 + 6; 1 block in her left pocket and 5 in her right: 6 = 1 + 5; and so on.)

9. Today, Lee has 12 blocks. Iris has 16 blocks. How many more blocks does Iris have?

10. There are 26 blocks on the floor. Then Lee picks up 11 of the blocks. How many blocks are on the floor now?

11. 15 blocks are in a box. Iris takes some out. Now there are 8 blocks in the box. How many did Iris take out?

12. 27 blocks are on the table. 22 of them are red and the rest are white. How many of the blocks are white?

SOLUTIONS and Problem Types for SET 6: Blocks

Set 6	OPEN SENTENCES	Solution	Classification (**not** important for students to learn); listed to show variety of problems.
1	17 + 4 = __	21	COMPARE: BIGGER UNKNOWN
2	5 + 3 = __ or 3 + 5 = __	8	JOIN – RESULT UNKNOWN
3	2 + 9 = __	11	PART-PART-WHOLE: TOTAL UNKNOWN
4	18 – 13 = __ or 13 + __ = 18	5	COMPARE: SMALLER UNKNOWN
5	19 + __ = 29 or 29 – 19 = __	10	JOIN – CHANGE UNKNOWN
6	__ - 21 = 14 or 14 + 21 = __	35	SEPARATE – START UNKNOWN
7	__ + 8 = 12 or 12 – 8 = __	4	JOIN – START UNKNOWN
8	6 = 0 + 6 = 6 + 0 6 = 1 + 5 = 5 + 1 6 = 2 + 4 = 4 + 2 6 – 3 + 3	See left	PART-PART-WHOLE: Both ADDENDS UNKNOWN
9	16 – 12 = __ or 12 + __ = 16	4	COMPARE: DIFFERENCE UNKNOWN
10	26 – 11 = __	15	SEPARATE – RESULT UNKNOWN
11	15 - __ = 8 or 15 – 8 = __	7	SEPARATE – CHANGE UNKNOWN
12	27 = 22 + __ or 27 – 22 = __	5	PART-PART-WHOLE: ADDEND UNKNOWN

SET 7: For each problem, write an open sentence (example: 3 + __ = 5) and solve. Look for key words. You can model the problem in many ways: draw a picture; use counters or base-ten blocks; make a number line, etc.

1. 21 bananas are on a tree. 17 of them are green and the rest are yellow. How many of the bananas are yellow?

2. Ben has 3 bananas. Kerr has 8 bananas. How many more bananas than Ben does Kerr have?

3. 11 bananas are in a bowl. Then Ben takes some out of the bowl. Now there are 2 bananas left in the bowl. How many did Ben take?

4. A bowl holds 18 bananas. Kerr takes out 13 of them. How many bananas are left in the bowl?

5. Some bananas are on a tree. Ben picks 19 of the bananas. Now there are 10 bananas left on the tree. How many were on the tree before?

6. A shelf in the store has 14 green bananas and 21 yellow bananas. What is the sum of green and yellow bananas?

7. Some bananas sit in a bowl. Kerr puts 4 more bananas in the bowl. Now there are 12 bananas in the bowl. How many were in the bowl to start?

8. Ben wants to put 9 bananas on plates as snacks for his friends. He has a blue plate and a red plate. What are all the ways he can place the bananas on the plates? (Example: 0 bananas on the blue plate and 9 on the red plate: $9 = 0 + 9$; 1 banana on the blue plate and 8 on the red: $9 = 1 + 8$.)

9. Lana has 4 more bananas than Kiley has. Kiley has 12 bananas. How many does Lana have?

10. There are 11 bananas on a store shelf. Then a worker puts more bananas on the shelf. Now there are 26 bananas on the shelf. How many did the worker put there?

11. There are 15 bananas on the big banana tree. There are 8 fewer bananas on the small banana tree than on the big tree. How many bananas are there on the small tree?

12. 22 bananas are in a bowl. Then 5 more bananas are put in the bowl. How many bananas are in the bowl now?

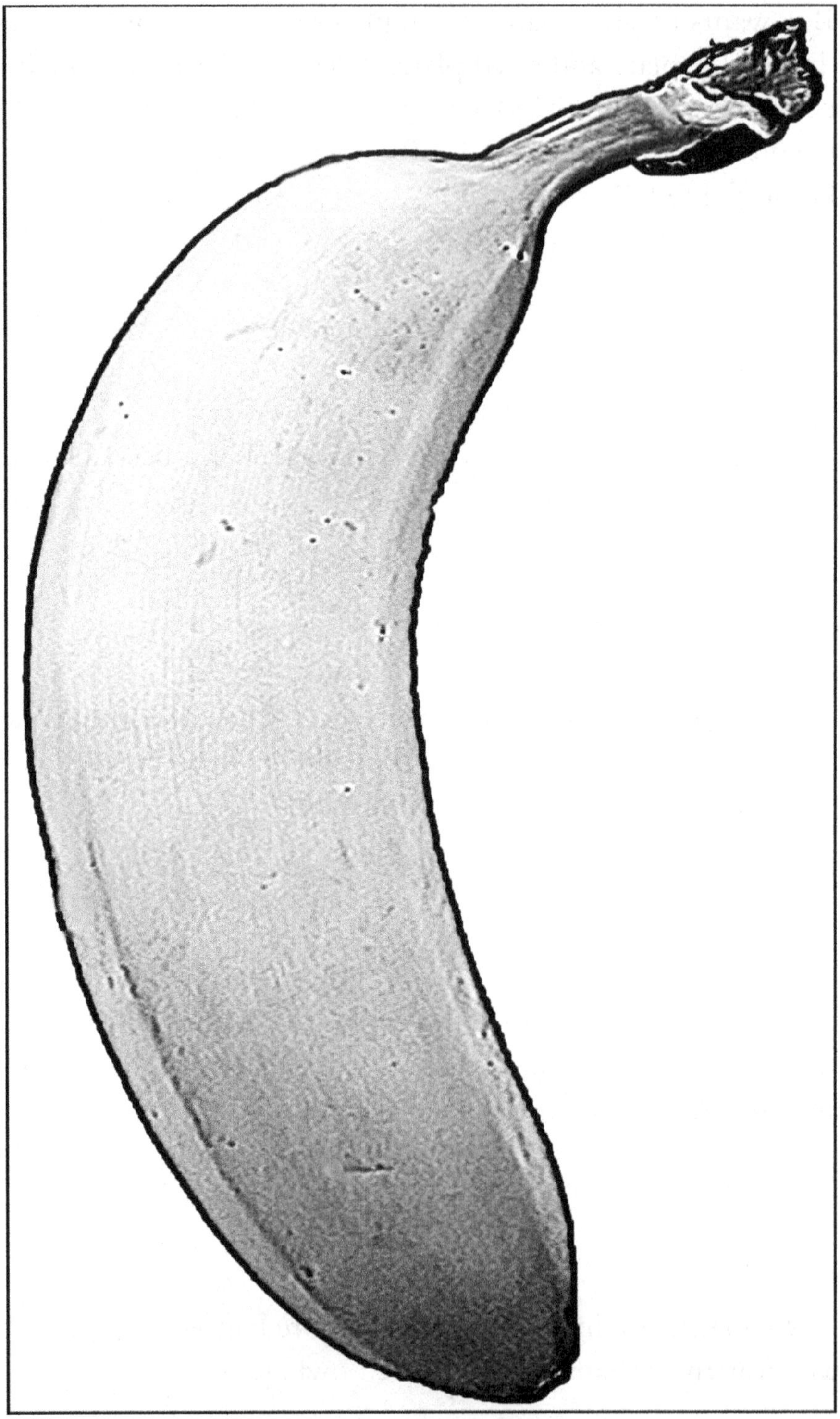

SOLUTIONS and Problem Types for SET 7: Bananas

Set 7	OPEN SENTENCES	Solution	Classification (**not** important for students to learn); listed to show variety of problems.
1	21 = 17 + __ or 21 – 17 = __	4	PART-PART-WHOLE: ADDEND UNKNOWN
2	8 – 3 = __ or 3 + __ = 8	5	COMPARE: DIFFERENCE UNKNOWN
3	11 - __ = 2 or 11 – 2 = __	9	SEPARATE – CHANGE UNKNOWN
4	18 – 13 = __	5	SEPARATE – RESULT UNKNOWN
5	__ - 19 = 10 or 10 + 19 = __	29	SEPARATE – START UNKNOWN
6	14 + 21 = __	35	PART-PART-WHOLE: TOTAL UNKNOWN
7	__ + 4 = 12 or 12 – 4 = __	8	JOIN – START UNKNOWN
8	9 = 0 + 9 = 9 + 0 9 = 1 + 8 = 8 + 1 9 = 2 + 7 = 7 + 2 9 = 3 + 6 = 6 + 3 9 = 4 + 5 = 5 + 4	See left	PART-PART-WHOLE: Both ADDENDS UNKNOWN
9	12 + 4 = __	16	COMPARE: BIGGER UNKNOWN
10	11 + __ = 26 or 26 – 11 = __	15	JOIN – CHANGE UNKNOWN
11	15 – 8 = __ or 8 + __ = 15	7	COMPARE: SMALLER UNKNOWN
12	22 + 5 = __ or 5 + 22 = __	27	JOIN – RESULT UNKNOWN

SET 8: For each problem, write an open sentence (example: 3 + __ = 5) and solve. Look for key words. You can model the problem in many ways: draw a picture; use counters or base-ten blocks; make a number line, etc.

1. On a river, 17 boats are floating. Some more boats are put into the river. Now there are 21 boats floating. How many boats were put into the river?

2. Manolo owns 5 more boats than Jackie owns. Jackie owns 3 boats. How many boats does Manolo own?

3. 11 boats float on the river. 9 of them are white and the rest are blue. How many of the boats are blue?

4. Some boats are at the dock. 13 more boats arrive at the dock. Now there are 18 boats at the dock. How many boats were at the dock before?

5. 19 green boats float in the lake, and 10 red boats float in the lake. How many boats are in the lake altogether?

6. Some boats are at the dock. 14 boats leave the dock. Now there are 21 boats at the dock. How many boats were there to begin with?

7. 12 boats need to be tied up at the dock. They can be tied up on the left side or on the right side. What are all the ways they can be tied up at the dock? (Example: 0 boats on the left side and 12 on the right side: $12 = 0 + 12$; 1 boat on the left and 11 boats on the right: $12 = 1 + 11$. List all the ways the 12 boats can be tied up at the dock.)

8. 9 boats float on the lake. Some of them are taken out of the lake. Now there are 6 boats left on the lake. How many were taken out?

9. 4 boats are tied up at the dock. Now 12 more boats come to the dock and are tied up there. How many boats are at the dock now?

10. There are 26 boats floating on the lake. On the pond, there are 15 fewer boats than there are on the lake. How many boats are on the pond?

11. At a dock, 15 boats are tied up. Now 7 boats leave. How many boats are still tied up at the dock?

12. There are 27 yellow boats on the lake and 22 orange boats on the lake. How many more yellow boats are there than orange boats?

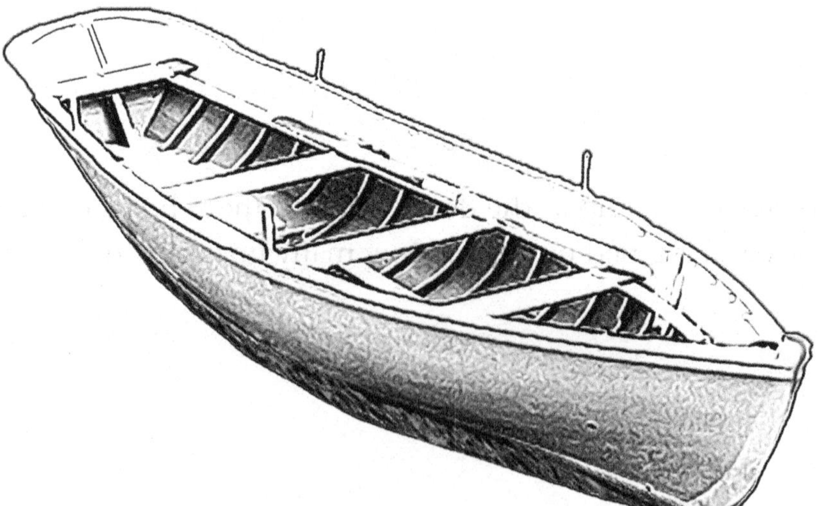

SOLUTIONS and Problem Types for SET 8: Boats

Set 8	OPEN SENTENCES	Solution	Classification (**not** important for students to learn); listed to show variety of problems.
1	17 + __ = 21 or 21 – 17 = __	4	JOIN – CHANGE UNKNOWN
2	3 + 5 = __	8	COMPARE: BIGGER UNKNOWN
3	11 = 9 + __ or 11 – 9 = __	2	PART-PART-WHOLE: ADDEND UNKNOWN
4	__ + 13 = 18 or 18 – 13 = __	5	JOIN – START UNKNOWN
5	19 + 10 = __	29	PART-PART-WHOLE: TOTAL UNKNOWN
6	__ - 14 = 21 or 21 + 14 = __	35	SEPARATE – START UNKNOWN
7	12 = 0 + 12 = 12 + 0 12 = 1 + 11 = 11 + 1 12 = 2 + 10 = 10 + 2 12 = 3 + 9 = 9 + 3 12 = 4 + 8 = 8 + 4 12 = 5 + 7 = 7 + 5 12 = 6 + 6	See left	PART-PART-WHOLE: Both ADDENDS UNKNOWN
8	9 - __ = 6 or 9 – 6 = __	3	SEPARATE – CHANGE UNKNOWN
9	4 + 12 = __ or 12 + 4 = __	16	JOIN – RESULT UNKNOWN
10	26 – 15 = __ or 15 + __ = 26	11	COMPARE: SMALLER UNKNOWN
11	15 – 7 = __	8	SEPARATE – RESULT UNKNOWN
12	27 – 22 = __ or 22 + __ = 27	5	COMPARE: DIFFERENCE UNKNOWN

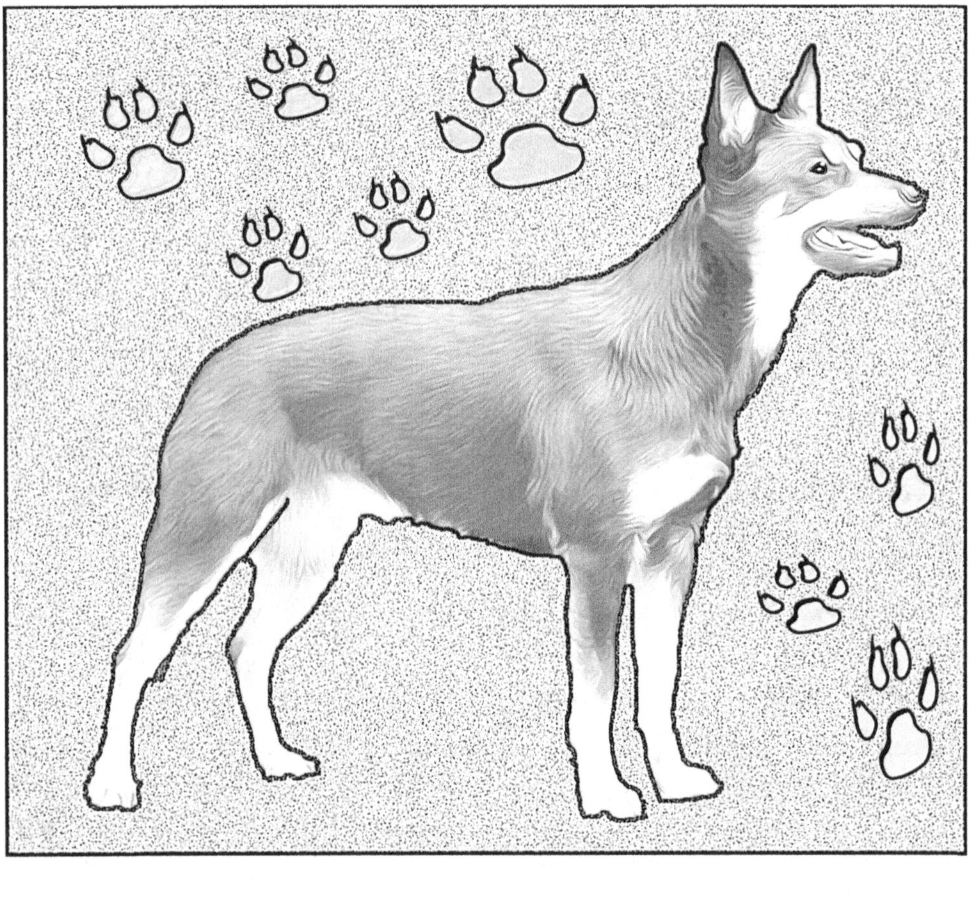

SET 9: For each problem, write an open sentence (example: 3 + __ = 5) and solve. Look for key words. You can model the problem in many ways: draw a picture; use counters or base-ten blocks; make a number line, etc.

1. There are 21 big dogs playing in the park. There are 17 fewer small dogs than big dogs playing there. How many small dogs are playing in the park?

2. In a yard, 8 dogs are playing. 3 of the dogs go inside the house. How many dogs are left in the yard?

3. Some dogs are in a yard. Then 2 of them go inside the house. Now there are 9 dogs in the yard. How many dogs were in the yard before?

4. There are 13 big dogs and 5 small dogs in the park. How many dogs are in the park?

5. In a park, some dogs are playing. 19 more dogs come to the parks. Now there are 29 dogs playing in the park. How many were there before?

6. 35 dogs are playing in a park. 21 of them are short dogs and the rest are tall dogs. How many are tall dogs?

7. Gerard has 4 more dogs than Faris has. Faris has 8 dogs. How many dogs does Gerard have?

8. 3 dogs are playing in a yard. Then 6 more dogs come into the yard. How many dogs are in the yard now?

9. In the old park, 4 dogs are playing. In the new park, 16 dogs are playing. How many more dogs are playing in the new park than in the old one?

10. In a park, 15 dogs are playing. Then some more dogs come to the park and join in the play. Now there are 26 dogs playing in the park. How many dogs joined in?

11. Faris has 8 dogs and wants to give them each one treat. There are crunchy treats and chewy treats. What are all the ways the dogs could have their treats? (Example: 0 dogs get crunchy treats and 8 get chewy treats: $8 = 0 + 8$; 1 dog gets a crunchy treat and 7 get chewy: $8 = 1 + 7$. List all.)

12. There are 27 dogs playing in the park. Then some of the dogs are taken home. Now there are 22 dogs in the park. How many were taken home?

SOLUTIONS and Problem Types for SET 9: Dogs

Set 9	OPEN SENTENCES	Solution	Classification (**not** important for students to learn); listed to show variety of problems.
1	21 – 17 = __ or 17 + __ = 21	4	COMPARE: SMALLER UNKNOWN
2	8 – 3 = __	5	SEPARATE – RESULT UNKNOWN
3	__ - 2 = 9 or 9 + 2 = __	11	SEPARATE – START UNKNOWN
4	13 + 5 = __	18	PART-PART-WHOLE: TOTAL UNKNOWN
5	__ + 19 = 29 or 29 – 19 = __	10	JOIN – START UNKNOWN
6	35 = 21 + __ or 35 – 21 = __	14	PART-PART-WHOLE: ADDEND UNKNOWN
7	8 + 4 = __	12	COMPARE: BIGGER UNKNOWN
8	3 + 6 = __ or 6 + 3 = __	9	JOIN – RESULT UNKNOWN
9	16 – 4 = __ or 4 + __ = 16	12	COMPARE: DIFFERENCE UNKNOWN
10	15 + __ = 26 or 26 – 15 = __	11	JOIN – CHANGE UNKNOWN
11	8 = 0 + 8 = 8 + 0 8 = 1 + 7 = 7 + 1 8 = 2 + 6 = 6 + 2 8 = 3 + 5 = 5 + 3 8 = 4 + 4	See left	PART-PART-WHOLE: Both ADDENDS UNKNOWN
12	27 - __ = 22 or 27 – 22 = __	5	SEPARATE – CHANGE UNKNOWN

A Wrinkle in Time
OLIVIA
KIDNEY
Charlotte's Web
Shawn and the Big Red Wagon
I Like Kites
New Home for Chris
Steinwort Hexengeschichten
SEE THE TRUCKS!
Birthday Bunnies!
MANY NEW FRIENDS
Der Räuber Hotzenplotz
Das kleine Gespenst
Astrid Lindgren
Jokes and more Jokes
Astrid Lindgren
Madita
Astrid Lindgren

SET 10: For each problem, write an open sentence (example: 3 + __ = 5) and solve. Look for key words. You can model the problem in many ways: draw a picture; use counters or base-ten blocks; make a number line, etc.

1. There are 17 books on the shelf. Then Tracy puts 4 more books on the shelf. How many books are there now?

2. Larkin has 5 books at school. Tracy has 8 books at school. How many more books than Larkin has, does Tracy have?

3. There are 11 books on the table. There are 9 fewer books on the desk than there are on the table. How many books are on the desk?

4. There are 18 books in Larkin's backpack. Then Larkin takes 5 books out of the backpack. How many books remain in the backpack?

5. On a shelf, there are 29 books. 19 of the book are thick and the rest are thin. How many of the books are thin?

6. Tracy has 14 more books at home than Larkin has. Larkin has 21 books at home. How many books does Tracy have at home?

7. There are 12 books on the desk. Then Tracy puts some of them on the shelf. Now there are 8 books on the desk. How many did Tracy put on the shelf?

8. Tracy has some books in a box. Then Tracy takes 6 of the books out of the box. Now there are 3 books left in the box. How many books were in the box to begin with?

9. Larkin wants to put 16 books on a shelf. The books can be put on the left end of the shelf or on the right end. What are all the ways the books can be put on the shelf? (Example: 0 books on the left and 16 on the right: $16 = 0 + 16$. List all the possible ways the books can be put on the shelf.)

10. On a desk there are 15 thick books and 11 thin books. How many books are on the desk in all?

11. There are 8 books in the car. Tracy brings some more books into the car. Now there are 15 books in the car. How many books did Tracy bring?

12. Larkin has some books. At a birthday party, Larkin is given 22 more books. Now Larkin has 27 books. How many books did Larkin have before the party?

books!
Chris & the Cat
Wow
DId You Ever See A SILLY BILLY?
Sam Sees a Super Sign
I LOVE Math!
Too Many Triangles
The Bunnies Find A Happy Home
*** IT'S COOL OUT HERE ***
Sukie hits the ball!
where is the smallest baby animal?
YOU AND YOUR STORY
Fun at the Beach

SOLUTIONS and Problem Types for SET 10: Books

Set 10	OPEN SENTENCES	Solution	Classification (**not** important for students to learn); listed to show variety of problems.
1	17 + 4 = __ or 4 + 17 = __	21	JOIN – RESULT UNKNOWN
2	8 – 5 = __ or 5 + __ = 8	3	COMPARE: DIFFERENCE UNKNOWN
3	11 – 9 = __ or 9 + __ = 11	2	COMPARE: SMALLER UNKNOWN
4	18 – 5 = __	13	SEPARATE – RESULT UNKNOWN
5	29 = 19 + __ or 29 – 19 = __	10	PART-PART-WHOLE: ADDEND UNKNOWN
6	21 + 14 = __	35	COMPARE: BIGGER UNKNOWN
7	12 - __ = 8 or 12 – 8 = __	4	SEPARATE – CHANGE UNKNOWN
8	__ - 6 = 3 or 3 + 6 = __	9	SEPARATE – START UNKNOWN
9	16 = 0 + 16 = 16 + 0 16 = 1 + 15 = 15 + 1 16 = 2 + 14 = 14 + 2 16 = 3 + 13 = 13 + 3 16 = 4 + 12 = 12 + 4 16 = 5 + 11 = 11 + 5 16 = 6 + 10 = 10 + 6 16 = 7 + 9 = 9 + 7 16 = 8 + 8	See left	PART-PART-WHOLE: Both ADDENDS UNKNOWN
10	15 + 11 = __	26	PART-PART-WHOLE: TOTAL UNKNOWN
11	8 + __ = 15 or 15 – 8 = __	7	JOIN – CHANGE UNKNOWN
12	__ + 22 = 27 or 27 – 22 = __	5	JOIN – START UNKNOWN

About the Author

K. Sherrerd Lowry, Ph.D., received the doctorate in Mathematics Education in 2006. Years of classroom teaching in advance of this doctoral degree work, combined with a three-year stint editing mathematics textbooks for a major publisher, demonstrate Dr. Lowry's interest in the topic of how children best learn math. Add a love of story and an abiding interest in the graphic arts, and the result is *LC4A and the Big Test* -- one of several planned children's math storybooks.

Check out upcoming offerings at the Duer Press website, http://duerpress.com/ (where a comments section is under construction.) Or, comment on the Amazon page for this paperback or for the original storybook version of this book, a color Kindle ebook, at http://www.amazon.com/LC4A-Big-Test-Storybook-Childrens-ebook/dp/B00FG0WKKM. Your feedback is welcome.

DUER PRESS

Made in the USA
Monee, IL
06 November 2026

68030312R00077